# Lecture Notes in Mathematics

Edited by A. Dold and B. Eckmann

Series: Australian National University, Canberra

Advisers: L. G. Kovács, B. H. Neumann, and M. F. Newman

546



Kurt Mahler

# Lectures on Transcendental Numbers

Edited and Completed by B. Diviš and W. J. Le Veque

Springer-Verlag
Berlin · Heidelberg · New York 1976

**Author**
Kurt Mahler
Department of Mathematics
Research School of Physical Sciences
Australian National University
Canberra, ACT 2600/Australia

**Editors**
B. Diviš †
W. J. Le Veque
Claremont Graduate School
Claremont, California/USA

**Library of Congress Cataloging in Publication Data**

Mahler, Kurt.
Lectures on transcendental numbers.

(Lecture notes in mathematics ; 546)
Bibliography: p.
Includes index.
1. Numbers, Transcendental. I. Divis, B.
II. LeVeque, William Judson. III. Title. IV. Series: Lecture notes in mathematics (Berlin) ; 546.
QA3.L28 no. 546 [QA247.5] 510'.8s [512'.73] 76-44348

AMS Subject Classifications (1970): 10F35

ISBN 3-540-07986-6 Springer-Verlag Berlin · Heidelberg · New York
ISBN 0-387-07986-6 Springer-Verlag New York · Heidelberg · Berlin

After the manuscript of this book had gone to the printers, our collaborator, Professor B. Diviš, died suddenly at the early age of 32, while attending a number theoretic meeting at Illinois State University. This is a great loss to his family, to mathematics, as well as to us.

K. Mahler

W.J. Le Veque

## PREFACE

The rather small list of modern books on transcendental numbers (Siegel 1949; Gelfond 1952; Schneider 1957; Lang 1962; Ramachandra 1969) has recently been enriched by four new ones (Stolarsky 1974; Waldschmidt 1974; Baker 1975; Masser 1975). Baker's book is particularly valuable because it contains an account of his powerful new method and a review of some of its applications.

The present book is derived from lectures given by Mahler during the last twenty years at different times and places, and edited and sometimes enlarged by Diviš and LeVeque. It aims at giving an account of some old and classical results and methods on transcendency and in particular to present in all its details the important Siegel-Shidlovski theory of the transcendency of a special class of entire functions (the Siegel $E$-functions) which satisfy linear differential equations with rational functions as coefficients. In the earlier books this theory was only sketched.

Chapter 1 discusses the existence and gives first examples of transcendental numbers and concludes with a rather general necessary and sufficient condition for transcendency. As it will be needed later on, Chapter 2 gives a short account of the theory of formal Laurent and power series in one variable, and it contains a simple test for the transcendency of such series over the field of rational functions. Chapter 3 begins then the study of the possible transcendency of the values of an analytic function at an algebraic point. A number of classical results, of which some go back to the time of Weierstrass, are proved in order to show that certain very plausible conjectures are in fact false. Thus, (i) there exist entire transcendental functions defined by power series with rational coefficients which, with all their derivatives, assume algebraic values at all algebraic points, and (ii) there also exist functions of the same kind which, say, are algebraic at the algebraic point $z = \sqrt{2}$ , but are transcendental at the algebraically conjugate point $z = -\sqrt{2}$ . It follows that the problem whether a given analytic function $f(z)$ is algebraic or transcendental at a given algebraic point $z = \alpha$ may be very deep and involved.

The next four chapters 4-7 deal then with Shidlovski's generalisation of Siegel's method of 1929 and 1949. This theory was obtained by Shidlovski in a number of papers that go back to 1954 and of which the most important ones date from 1959 to 1962. We try to bring out the algebraic basis of this theory by using both formal series and simple facts from algebraic geometry.

In Chapter 4, $K$ is any field of characteristic $0$ , $c$ is an element of $K$ , and $z$ is an indeterminate; $K[z]$, $K(z)$, and $K\langle z-c\rangle$ are then the ring of polynomials, the field of rational functions, and the field of formal Laurent series

$$f = \sum_{h=\eta}^{\infty} f_h (z-c)^h$$

in $z$, all with coefficients in $K$; in $f$, $\eta$ may be any integer. We denote the degree of a polynomial $p \neq 0$ in $K[z]$ by $\partial(p)$; and if the Laurent series $f$ is distinct from $0$ and so, without loss of generality, $f_\eta \neq 0$, the order of $f$ is defined by

$$\text{ord}\, f = \eta .$$

We are concerned with the formal theory of systems of homogeneous linear differential equations

$$Q : w_h' = \sum_{h=1}^{m} q_{hk} w_k \qquad (h = 1,2,\ldots,m) .$$

Here $m \geq 1$, and the coefficients $q_{hk}$ are $m^2$ arbitrary elements of $K(z)$; let $\kappa$ be their least common denominator. The system $Q$ is said to be regular (that is at $c$) if $\kappa$ does not vanish at the point $z = c$.

Denote by $V_Q$ the set of all vector solutions

$$\mathbf{w} = \begin{pmatrix} w_1 \\ \vdots \\ w_m \end{pmatrix}$$

of $Q$ with components $w_1, \ldots, w_m$ in $K\langle z-c\rangle$. This set $V_Q$ is a linear vector space over $K\langle z-c\rangle$, say of dimension $M$ where of course $0 \leq M \leq m$. As can be proved, $M$ is also the dimension of $V_Q$ over the constant field $K$. If $Q$ is regular, then $M = m$.

Next let

$$\mathbf{f} = \begin{pmatrix} f_1 \\ \vdots \\ f_m \end{pmatrix}$$

be any special element of $V_Q$ distinct from the zero vector $\mathbf{0}$. The rank $\rho$ of $\mathbf{f}$ is defined as the largest number of components $f_1, \ldots, f_m$ of $\mathbf{f}$ that are linearly independent over $K(z)$.

A first theorem by Shidlovski deals with the case when $Q$ is regular and $\mathbf{f}$ is of rank $\rho < m$. If, without loss of generality, the first $\rho$ components

$f_1, \ldots, f_\rho$ are linearly independent over $K(z)$, denote by $\mathbf{f}^*$ the $\rho$-vector

$$\mathbf{f}^* = \begin{pmatrix} f_1 \\ \vdots \\ f_\rho \end{pmatrix} .$$

Then the following result holds.

(A): *The components of* $\mathbf{f}^*$ *satisfy a regular system of homogeneous linear differential equations*

$$Q^* : w_h' = \sum_{k=1}^{\rho} q_{hk}^* w_k \qquad (h = 1,2,\ldots,\rho)$$

*of only* $\rho$ *equations with coefficients* $q_{hk}^*$ *in* $K(z)$ .

Assume next that the system $Q$ is not necessarily regular, but that $\mathbf{f}$ is of maximal rank $m$. Let $p_1 = p_{11}, \ldots, p_m = p_{1m}$ be any $m$ polynomials in $K[z]$ not all zero. For any element $\mathbf{w}$ of $V_Q$ we form the linear form

$$\lambda(\mathbf{w}) = p_1 w_1 + \ldots + p_m w_m$$

in its components and for such forms define a linear operator $D$ by

$$D\lambda(\mathbf{w}) = \sum_{h=1}^{m} \kappa \frac{dp_h}{dz} w_h + \sum_{h=1}^{m} \sum_{j=1}^{m} \kappa q_{hj} \cdot p_j w_h .$$

Also $D\lambda(\mathbf{w})$ is a linear form in the components of $\mathbf{w}$ with coefficients in $K[z]$.

The formulae

$$\lambda_1(\mathbf{w}) = \lambda(\mathbf{w}) , \quad \lambda_{h+1}(\mathbf{w}) = D\lambda_h(\mathbf{w}) \qquad (h = 1,2,3,\ldots)$$

thus generate an infinite sequence of linear forms

$$\lambda_h(\mathbf{w}) = p_{h1} w_1 + \ldots + p_{hm} w_m \qquad (h = 1,2,3,\ldots)$$

in the components of $\mathbf{w}$ with coefficients in $K[z]$.

Shidlovski proves for these forms the following basic theorem.

(B): *To every element* $\mathbf{f}$ *of* $V_Q$ *of rank* $m$ *there exists a positive integer* $C$ *with the following property.*

*If the polynomials* $p_1, \ldots, p_m$ *satisfy the inequality*

$$\operatorname{ord} \lambda(\mathbf{f}) - (m-1)\max\left(\partial(p_1),\ldots,\partial(p_m)\right) \geq C ,$$

*then the determinant*

$$\begin{vmatrix} p_{11}, & \ldots, & p_{1m} \\ \vdots & & \vdots \\ p_{m1}, & \ldots, & p_{mm} \end{vmatrix}$$

*is not identically zero.*

Using a method of Siegel, we can deduce from this the following third theorem.

**(C):** *Let* $\alpha$ *be an element of* $K$ *such that* $\alpha \neq c$ *and* $\kappa(\alpha) \neq 0$ ; *let* $\phi$ *be a constant satisfying* $0 < \phi < 1$ ; *and let* $\mathbf{f}$ *be an element of* $V_Q$ *of rank* $m$ . *There exist two positive integers* $n_0$ *and* $n_1$ *with the following property.*

*If* $n \geq n_0$ , *and if the polynomials* $p_1, \ldots, p_m$ *in* $\lambda(\mathbf{w})$ *are not all zero and such that*

$$\max(\partial(p_1), \ldots, \partial(p_m)) \leq n-1 , \quad \text{ord}\ \lambda(\mathbf{f}) \geq mn-[\phi n]-1 ,$$

*then* $m$ *suffixes* $h_1, \ldots, h_m$ *satisfying*

$$1 \leq h_1 < h_2 < \ldots < h_m \leq [\phi n]+m+n_1$$

*can be chosen such that the determinant*

$$\begin{vmatrix} p_{h_1 1}(\alpha), & \ldots, & p_{h_1 m}(\alpha) \\ \vdots & & \vdots \\ p_{h_m 1}(\alpha), & \ldots, & p_{h_m m}(\alpha) \end{vmatrix} = P(\alpha) \quad say,$$

*does not vanish.*

These purely algebraic theorems are in Chapter 5 combined with the theory of Diophantine approximations; the method is that of Siegel's paper of 1929. No longer are we considering formal Laurent series in an indeterminate $z$ , but entire functions in the complex variable $z$ . Further $c$ is put equal to $0$ .

The field $K$ is specialised to be an algebraic number field, say of degree $N$ over the rational field $\mathbf{Q}$ . Every number $\alpha$ in $K$ has $N$ algebraic conjugates

$$\alpha^{(0)} = \alpha, \alpha^{(1)}, \ldots, \alpha^{(N-1)} ,$$

and as usual we put

$$\overline{|\alpha|} = \max(|\alpha^{(0)}|, |\alpha^{(1)}|, \ldots, |\alpha^{(N-1)}|) ,$$

so that for any two elements $\alpha$ and $\beta$ of $K$ ,

$$\overline{|\alpha+\beta|} \leq \overline{|\alpha|} + \overline{|\beta|} , \quad \overline{|\alpha\beta|} \leq \overline{|\alpha|}\,\overline{|\beta|} .$$

An $E$-function is defined as a power series

$$f(z) = \sum_{\nu=0}^{\infty} f_\nu \frac{z^\nu}{\nu!}$$

with the following two properties of which the first one implies that $f(z)$ is an entire function.

(1): The coefficients $f_\nu$ lie in $K$, and for every constant $\varepsilon > 0$, as $\nu \to \infty$,

$$\overline{|f_\nu|} = O(\nu^{\varepsilon\nu}) .$$

(2): Let $d_\nu$ be the smallest positive integer such that all the products

$$d_\nu f_0,\ d_\nu f_1,\ \ldots,\ d_\nu f_\nu$$

are algebraic integers. Then for every constant $\varepsilon > 0$, as $\nu \to \infty$,

$$d_\nu = O(\nu^{\varepsilon\nu}) .$$

From this definition, Dirichlet's principle (the Schubfachprinzip) leads to the following existence result.

(D): *Let* $m \ge 2$ ; *let* $f_1(z), \ldots, f_m(z)$ *be* $E$*-functions over* $K$ ; *let* $\varepsilon$ *and* $\phi$ *be constants satisfying* $0 < \varepsilon < 1$, $0 < \phi < 1$ ; *let* $n$ *be a sufficiently large positive integer; and let* $p = mn-[\phi n]-1$ . *Then there exist* $m$ *polynomials*

$$p_h(z) = \sum_{\mu=0}^{n-1} G_{h\mu} z^\mu \qquad (h = 1,2,\ldots,m) ,$$

*with integral coefficients in* $K$ *not all zero, such that the linear form*

$$\lambda(\mathbf{f}(z)) = \sum_{h=1}^{m} p_h(z) f_h(z)$$

*can be expressed as a power series*

$$\lambda(\mathbf{f}(z)) = \sum_{\nu=0}^{\infty} a_\nu \frac{z^\nu}{\nu!}$$

*where, firstly,*

$$a_0 = a_1 = \ldots = a_{p-1} = 0 ,$$

*and, secondly, as* $\nu \to \infty$,

$$\max_{h,\mu} \overline{|G_{h\mu}|} = O(n^{(1+\varepsilon)n}) \quad \text{and} \quad \overline{|a_\nu|} = n^n O(\nu^{\varepsilon\nu}) .$$

Now again let

$$Q : w_h' = \sum_{k=1}^{m} q_{hk} w_k \qquad (h = 1,2,\dots,m)$$

be a system of $m$ ($\geq 2$) homogeneous linear differential equations where the coefficients $q_{hk} = q_{hk}(z)$ lie in $K(z)$ and their least common denominator $\kappa = \kappa(z)$ is in $K[z]$ .

Assume that the vector

$$\mathbf{f}(z) = \begin{pmatrix} f_1(z) \\ \vdots \\ f_m(z) \end{pmatrix}$$

derived from the $m$ $E$-functions in (D) is an element of $V_Q$ . Then, with the polynomials $p_1(z), \dots, p_m(z)$ in (D), put

$$\lambda(\mathbf{w}) = \sum_{k=1}^{m} p_k(z) w_k .$$

On applying to $\lambda(\mathbf{w})$ repeatedly the differential operator $D$ of $Q$ , we obtain again a sequence of linear forms

$$\lambda_h(\mathbf{w}) = \sum_{k=1}^{m} p_{hk}(z) w_k \qquad (h = 1,2,3,\dots)$$

with coefficients $p_{hk}(z)$ in $K[z]$ .

Now let $\alpha$ be a number in $K$ satisfying $\alpha \neq 0$ and $\kappa(\alpha) \neq 0$ . As before, denote by $\rho$ the largest number of components of $\mathbf{f}(z)$ that are linearly independent over $K(z)$ , and similarly let $\rho(\alpha)$ be the largest number of components of $\mathbf{f}(\alpha)$ that are linearly independent over $K$ . It is obvious that $\rho(\alpha) \leq \rho$ , and the problem arises of also establishing a lower estimate for $\rho(\alpha)$ . This can be done by means of Siegel's method and leads to the following results.

First assume that $\rho$ has its maximum value $m$ . By the construction (D), the hypothesis of (C) is satisfied as soon as $n$ is sufficiently large. Hence it is again possible to select $m$ distinct suffixes $h_1, \dots, h_m$ not greater than $[\phi n] + m + n_1$ for which the determinant $P(\alpha)$ in (C) is distinct from zero. By means of simple estimates, Siegel's method leads to the inequality

$$\rho(\alpha) \geq \frac{\sigma(K)m}{N} ,$$

where $\sigma(K) = 1$ if $K$ is a real number field, and $\sigma(K) = 2$ if $K$ is non-real.

The restriction $\rho = m$ under which this estimate has been proved can finally also be removed by means of Shidlovski's theorem (A), and it follows from the last formula that always

(E): $$\rho(\alpha) \geq \frac{\sigma(K)\rho}{N} .$$

In particular, $\rho(\alpha) = \rho$ when $K$ is the rational number field or any imaginary quadratic field.

The beginning of Chapter 6 is again algebraic and is independent of Chapters 4 and 5.

Let $L$ be any field of characteristic 0 and $L^*$ any extension field of $L$. As usual, a finite set of elements of $L^*$ is called (algebraically) independent over $L$ if it does not satisfy any non-trivial algebraic equation with coefficients in $L$; and the set is said to be (algebraically) $H$-independent over $L$ if it does not satisfy any non-trivial homogeneous algebraic equations with coefficients in $L$.

Let then $x_1, \ldots, x_n$, where $n \geq 2$, be fixed elements of $L^*$ not all zero of which certain $D + 1$, but no more, are $H$-independent over $L$; and let $w_1, \ldots, w_n$ be an equal number of independent indeterminates. For every positive integer $t$ denote by $V(t)$ the set of all homogeneous polynomials $P = P(w_1, \ldots, w_n)$ in $L[w_1, \ldots, w_n]$ of dimension $t$, and by $S(t)$ the subset of those polynomials $P$ in $V(t)$ for which

$$P(x_1, \ldots, x_n) = 0 .$$

A polynomial $P$ in $V(t)$ belongs to $S(t)$ if and only if the set of its coefficients satisfies a certain system $\Sigma(t)$ of, say, exactly $h(t)$ linearly independent homogeneous linear equations with coefficients in $L$.

We show now that

(F): *There exist two positive integral constants* $a$ *and* $c$ *such that for all* $t$,

$$\binom{t+D}{D} \leq h(t) \leq a\binom{ct+D}{D} .$$

(This is the weakened form of a very special case of a theorem by Hilbert (1890) on polynomial ideals.)

From here on, the hypothesis is again the same as for theorem (E). Denote by $D + 1$ the maximum number of $E$-functions $f_1(z), \ldots, f_m(z)$ that are $H$-independent over $K(z)$, and similarly by $D(\alpha) + 1$ the maximum number of function values $f_1(\alpha), \ldots, f_m(\alpha)$ that are $H$-independent over $K$. Then the system $Q$ of $m$ homogeneous linear differential equations for $f_1(z), \ldots, f_m(z)$ implies for every positive integer $t$ a new system $Q(t)$ of $\tau = \binom{m+t-1}{m-1}$ homogeneous linear differential equations for the $\tau$ products

$$F_{(h)}(z) = f_1(z)^{h_1} \ldots f_m(z)^{h_m}\text{, where } h_1 \geq 0, \ldots, h_m \geq 0 \text{ , } h_1 + \ldots + h_m = t \text{ .}$$

Here the coefficients of $Q(t)$ lie again in $K(z)$ , and their least common denominator is a power of $\kappa$ ; further the new functions $F_{(h)}(z)$ are again $E$-functions.

Denote by $\rho(t)$ the maximum number of these $E$-functions $F_{(h)}(z)$ that are linearly independent over $K(z)$ , and similarly by $\rho(t;\alpha)$ the maximum number of function values $F_{(h)}(\alpha)$ that are linearly independent over $K$ . It follows immediately from (E) that

$$\mu(t;\alpha) \geq \frac{\sigma(K)\rho(t)}{N} \text{ .}$$

On the other hand, it can be deduced from (F) that

$$\rho(t) \geq \binom{t+D}{D} \quad \text{and} \quad \rho(t;\alpha) \leq a\binom{ct+D(\alpha)}{D(\alpha)} \text{ .}$$

Hence, as $\sigma(K) \geq 1$ ,

$$Na\binom{ct+D(\alpha)}{D(\alpha)} \geq \binom{t+D}{P} \text{ .}$$

Here the left-hand side is a polynomial in $t$ of the exact degree $D(\alpha)$ , and the right-hand side is one of the exact degree $D$ . On allowing $t$ to tend to infinity, it follows then that

$$D(\alpha) \geq D \text{ .}$$

On the other hand, one can easily show that $D(\alpha) \leq D$ . Hence

(G): $$D(\alpha) = D \text{ .}$$

Simple additional considerations lead from here to the two general theorems of Shidlovski, where $\mathbb{C}$ is the complex number field.

(I): *Let* $m \geq 2$ *; let*

$$Q : w_h' = \sum_{k=1}^{m} q_{hk}w_k \qquad (h = 1,2,\ldots,m)$$

> *be a system of homogeneous linear differential equations with coefficients in* $\mathbb{C}(z)$ *; and let* $\mathbf{f}(z)$ *be a vector solution of* $Q$ *the components of which are E-functions. Further let* $\alpha \neq 0$ *be any algebraic number at which none of the rational functions* $q_{hk}$ *has a pole.*
>
> *Then the largest number of functions* $f_1(z), \ldots, f_m(z)$ *that are*

*algebraically H-independent over* $\mathbb{C}(z)$ *is equal to the largest number of function values* $f_1(\alpha), \ldots, f_m(\alpha)$ *that are algebraically H-independent over* $\mathbb{Q}$ .

(II): *Let*

$$Q^* : w_h' = q_{h0} + \sum_{k=1}^{m} q_{hk} w_k \qquad (h = 1,2,\ldots,m)$$

*be a system of homogeneous or inhomogeneous linear differential equations with coefficients in* $\mathbb{C}(z)$ *; and let* $\mathbf{f}(z)$ *be a vector solution of* $Q^*$ *the components of which are E-functions. Further let* $\alpha \neq 0$ *be any algebraic number at which none of the rational functions* $q_{h0}$ *and* $q_{hk}$ *has a pole.*

*Then the largest number of functions* $f_1(z), \ldots, f_m(z)$ *that are algebraically independent over* $\mathbb{C}(z)$ *is equal to the largest number of function values* $f_1(\alpha), \ldots, f_m(\alpha)$ *that are algebraically independent over* $\mathbb{Q}$ .

In the remainder of Chapter 6 a number of consequences of these theorems are considered which have an interest in themselves.

It is unsatisfactory that these theorems can be proved so far only for the very special class of *E*-functions. Siegel also introduced a more general class of functions which need not be entire; but for these *G*-functions only much weaker and less general results can as yet be established.

In Chapter 7, we apply Shidlovski's theorems to a number of special functions, and in particular prove Lindemann's theorem and some results on Bessel functions. More such applications can be found in the recent literature.

Chapter 8 deals with a little known theorem by J. Popken (1935) on the coefficients of power series that satisfy algebraic differential equations. In certain rather special cases this theorem allows one to prove that of a certain finite set of coefficients at least one is transcendental.

The appendix brings together a collection of old proofs of the transcendency of $e$ and $\pi$ and of Lindemann's general theorem on the exponential functions. We have tried to explain the interconnection of these proofs and to bring out their differences. All these proofs are based on the formulae already introduced by Hermite in his classical paper on the transcendency of $e$ .

B. Diviš

W.J. LeVeque

K. Mahler

## CONTENTS

# CHAPTER 1

## EXISTENCE AND FIRST PROPERTIES OF TRANSCENDENTAL NUMBERS

1. The numbers we shall be concerned with will always be real or complex, and our problem will be to investigate whether such numbers are algebraic or transcendental.

Here a number $\xi$ is said to be *algebraic* if it satisfies at least one algebraic equation

$$a_0 + a_1x + \ldots + a_mx^m = 0 \qquad (m \geq 1,\ a_m \neq 0)$$

with rational coefficients. As the coefficients may be multiplied with their least common denominator, $\xi$ satisfies then also such an algebraic equation where the coefficients are integers (i.e. rational integers) and are relatively prime. If $\xi$ does not satisfy any such algebraic equation, it is called *transcendental*.

That algebraic numbers exist is easily proved. For let $p$, $q$, and $r \neq 0$ be arbitrary integers. Then the rational number $\rho = p/r$ satisfies the linear equation $rx - p = 0$, and the complex number $\sigma = (p+qi)/r$ satisfies the quadratic equation $r^2x^2 - 2prx + p^2 + q^2 = 0$. Thus both $\rho$ and $\sigma$ are algebraic. It is also evident that the set of all $\rho$ is dense on the real axis, and the set of all $\sigma$ is dense in the complex plane. This implies that the set of all real algebraic numbers likewise is dense on the real axis, and the set of all complex algebraic numbers is dense in the complex plane.

The set $A$ of all algebraic numbers is known to form a field. Hence, if $\xi$ is transcendental, and if $\alpha \neq 0$ is any number in $A$, also $\alpha\xi$ is transcendental. Thus, if there exists at least one real transcendental number, then the real transcendental numbers are dense on the real axis; and if there exists at least one complex transcendental number, then the complex transcendental numbers are dense in the complex plane.

It is, however, not at all obvious that there should exist transcendental numbers, and this was proved only in 1844 by J. Liouville. His method will be explained later in this chapter in a more general context. We begin with the much simpler existence proof that is due to G. Cantor (1874).

2. Throughout these lectures, the following notations will be used.

Let

$$a(x) = a_0 + a_1 x + \dots + a_m x^m$$

be any polynomial with real or complex coefficients. The two quantities

$$H(a) = \max(|a_0|, |a_1|, \dots, |a_m|) \quad \text{and} \quad L(a) = |a_0| + |a_1| + \dots + |a_m|$$

are called the *height* and the *length* of $a(x)$, respectively; both may serve as measures for the size of the coefficients of the polynomial. Although the height is more frequently used, the length has the advantage of satisfying the simple inequalities

(1): $$L(a+b) \le L(a) + L(b), \quad L(ab) \le L(a)L(b)$$

where $a(x)$ and $b(x)$ may be any two polynomials.

If $a(x)$ has the exact degree $m$ (thus if $a_m \ne 0$), we write

$$\partial(a) = m .$$

For the zero polynomial $0$ we use the convention of putting

$$\partial(0) = -\infty$$

where the symbol $-\infty$ is considered as smaller than any finite integer. Then

(2): $$\partial(a+b) \le \max\left(\partial(a), \partial(b)\right), \quad \partial(ab) = \partial(a) + \partial(b) .$$

It is also convenient to introduce the notations

$$\Lambda(a) = 2^{\partial(a)} L(a) \quad \text{and more generally} \quad \Lambda_C(a) = C^{\partial(a)} L(a) ,$$

where $C$, in the second expression, may be any real constant greater than $1$. From this definition, $\Lambda(a)$ is always an even positive integer if $a$ has integral coefficients and positive degree.

3. A polynomial $a(x)$ with integral coefficients is called *primitive* if these coefficients are relatively prime; and it is said to be *normed* if the highest coefficient $a_m$ is positive. The same notation is also used for the corresponding algebraic equation $a(x) = 0$.

Let $\xi$ be any algebraic number. Then $\xi$ satisfies infinitely many different primitive normed equations $a(x) = 0$. Amongst these equations there is one,

$$A(x|\xi) \equiv A_0 + A_1 x + \dots + A_M x^M = 0 \qquad (A_M > 0)$$

say, which is of lowest degree, $\partial(A) = M$, and this equation is unique. The polynomial $A(x|\xi)$ is the *minimal polynomial* for $\xi$. It has the important property of being *irreducible* over the rational field; thus it cannot be written as the product of two polynomials of positive degrees with rational coefficients.

We use the notations

$$\partial^0(\xi) = \partial(A) = M \;, \quad H^0(\xi) = H(A) \;, \quad L^0(\xi) = L(A)$$

for the *degree* $M$ of $\xi$, the *height* of $\xi$, and the *length* of $\xi$, respectively. Here the superscript 0 has been added to distinguish these quantities from the degree, the height, and the length of the constant polynomial $\xi$.

Since the equation $A(x|\xi) = 0$ for $\xi$ has the degree $M$, it has $M$ roots,

(3): $$\xi^{(0)},\ \xi^{(1)},\ \ldots,\ \xi^{(M-1)}$$

say. These roots are distinct real or complex numbers, and one of them is equal to $\xi$; let the numbering always be such that

$$\xi = \xi^{(0)} \;.$$

The $M$ roots (3) are the *algebraic conjugates* of $\xi$. If $\xi$ is transcendental, $\partial^0(\xi)$ is given the improper value

$$\partial^0(\xi) = \infty \;,$$

where $\infty$ is greater than any finite integer.

4. Cantor's existence proof consists in showing that the set $\mathsf{A}$ of all algebraic numbers is *countable*, but that there exists a set $B$ of real numbers that is *not countable*. The difference set $B - \mathsf{A} = B - (\mathsf{A} \cap B)$ cannot then be empty, and so $B$ contains elements that are not in $\mathsf{A}$, and hence are transcendental.

That $\mathsf{A}$ is countable is easily proved. Denote by $\underline{\underline{Q}}$ the set of all distinct irreducible primitive normed polynomials, of positive degrees and with integral coefficients. Then every element of $\mathsf{A}$ is a zero of just one polynomial in $\underline{\underline{Q}}$, and conversely, the zeros of each polynomial in $\underline{\underline{Q}}$ are elements of $\mathsf{A}$.

Now let $r$ run over all even integers. For each such $r$, there exist only finitely many polynomials in $\underline{\underline{Q}}$ for which

$$\Lambda(A) = r \;.$$

Hence on writing down successively first the polynomials in $\underline{\underline{Q}}$ for which $\Lambda(A) = 2$, then those for which $\Lambda(A) = 4$, then those for which $\Lambda(A) = 6$, etc., we obtain an enumeration

$$A_1(x),\ A_2(x),\ A_3(x),\ \ldots$$

of all the distinct polynomials in $\underline{\underline{Q}}$.

Now write down successively the zeros of $A_1(x)$, then those of $A_2(x)$, then those of $A_3(x)$, etc. We evidently obtain in this way an enumeration

$$\xi_1,\ \xi_2,\ \xi_3,\ \ldots$$

of all the distinct elements of A . This proves that A is countable.

Naturally not only A itself, but also all its subsets are countable. In particular, the sets of all real algebraic numbers, of all real algebraic numbers in the interval [0, 1] , and of all complex algebraic numbers inside the unit circle $|z| < 1$ , are all countable.

5. Next let $B$ be the set of those real numbers $\xi$ that can be written as infinite continued fractions

$$\xi = 0 + \frac{1}{g_1+}\frac{1}{g_2+}\frac{1}{g_3+}\cdots = [0, g_1, g_2, g_3, \ldots]$$

where each of the partial quotients $g_1, g_2, g_3, \ldots$ may assume only the two values 1 or 2 . Then $B$ is a subset of the set of all real irrational numbers in the interval [0, 1] , and we assert that *the set* $B$ *is not countable*. For otherwise its elements could be arranged in the form of a sequence $\xi_1, \xi_2, \xi_3, \ldots$ , where, say, $\xi_r$ is the continued fraction

(4): $$\xi_r = [0, g_1^{(r)}, g_2^{(r)}, g_3^{(r)}, \ldots] \qquad (r = 1,2,3,\ldots) .$$

But then the new continued fraction

$$\eta = [0, 3-g_1^{(1)}, 3-g_2^{(2)}, 3-g_3^{(3)}, \ldots]$$

would be distinct from all the continued fractions (4) and would still be an element of $B$ . Now $\eta$ would also be distinct from all the numbers $\xi_1, \xi_2, \xi_3, \ldots$ , and so a contradiction would arise.

Since $B$ is not countable, the remark at the beginning of §4 shows that $B$ *contains numbers that are transcendental*. In fact, rather more is true. The difference set $B - A$ consisting of all elements of $B$ that are transcendental is *not countable*. For $B = (A \cap B) \cup (B - A)$ where $A \cap B$ is countable.

If $\xi$ is in $B$ , it can be proved that

$$\left|\xi - \frac{p}{q}\right| \geq \frac{1}{(2+\sqrt{3})q^2}$$

for all pairs of integers $p$ and $q > 0$ . On applying this result in particular to the transcendental elements of $B$ , we see that there are real transcendental numbers that cannot be approximated well by rational numbers. We shall later find real transcendental numbers that, on the contrary, have exceedingly good rational approximations.

This proof for the existence of transcendental elements of $B$ is only an existence proof. Maillet, in his book of 1906, gave instead an effective construction of transcendental elements of this set, but his proof is more involved.

Since there do exist *real* transcendental numbers, the remark in §1 shows that there are also *non-real complex* transcendental numbers. Again the set of all complex transcendental numbers is *not* countable.

6. Having established the existence of transcendental numbers, we next should try to find simple necessary and sufficient conditions for transcendency. Before doing so, however, we first study some properties of polynomials and of quadratic forms that have a certain interest in themselves.

As before let

$$a(x) = a_0 + a_1x + \ldots + a_mx^m$$

be a polynomial with arbitrary complex coefficients. We have already introduced the degree $\partial(a)$, the height $H(a)$, and the length $L(a)$ of $a(x)$, as well as the expressions $\Lambda(a)$ and $\Lambda_C(a)$, and we have noted that $L(a)$ has the properties (1) of §2.

It is useful to study also the function of the coefficients of $a(x)$ defined by

$$M(a) = \exp\left(\int_0^1 \log|a(e^{2\pi it})|dt\right) .$$

It is clear that $M(a) > 0$ if $a(x) \not\equiv 0$, and $M(0) = 0$ if $a(x) \equiv 0$. We call $M(a)$ the *measure* of $a(x)$; it is equal to the ordinary absolute value when the polynomial reduces to a constant.

The interest in $M(a)$ lies in the multiplicative law

(5): $$M(ab) = M(a)M(b)$$

which follows immediately from the definition.

There is a simple representation of $M(a)$, in terms of the zeros of $a(x)$, which is a special case of the well-known formula of Jensen. We shall give here an independent proof for this representation.

7. For this purpose let $r \geq 0$ and $\tau$ be any real numbers. The linear polynomial $x - re^{2\pi i\tau}$ has the measure

$$M(x-re^{2\pi i\tau}) = \exp\left(\int_0^1 \log|e^{2\pi it}-re^{2\pi i\tau}|dt\right) = \exp\left(\int_0^1 \log|e^{2\pi i(t-\tau)}-r|dt\right) .$$

Here the integrand has the period 1 in $t$, so that, on putting $t-\tau = s$, also

$$M(x-re^{2\pi i\tau}) = \exp\left(\int_0^1 \log|e^{2\pi is}-r|\,ds\right) = M(x-r)\ .$$

It follows that

(6): $$M(x-\alpha) = M(x-|\alpha|) \qquad \text{for all complex } \alpha\ .$$

Therefore, if $n$ is any positive integer,

$$\{M(x-\alpha)\}^n = \prod_{k=0}^{n-1} M(x-\alpha e^{2\pi ik/n}) = \exp\left(\sum_{k=0}^{n-1}\int_0^1 \log|e^{2\pi it}-\alpha e^{2\pi ik/n}|\,dt\right) =$$

$$= \exp\left(\int_0^1 \log|e^{2\pi itn}-\alpha^n|\,dt\right)\ .$$

Putting $nt = u$ , we obtain

$$\{M(x-\alpha)\}^n = \exp\left(\int_0^n \log|e^{2\pi iu}-\alpha^n|\,\frac{du}{n}\right) = \exp\left(\int_0^1 \log|e^{2\pi iu}-\alpha^n|\,du\right)\ ,$$

and hence

(7): $$M(x-\alpha) = M(x-\alpha^n)^{\frac{1}{n}}\ ,$$

where the $n$th root is taken with the positive value.

From (6) and (7), with $n \geq 2$ , it follows immediately that $M(x-1) = 1$ , hence that

(a): $$M(x-\alpha) = 1 \quad \text{if} \quad |\alpha| = 1\ .$$

Next assume that $|\alpha| < 1$ , and choose $n$ so large that

$$\tfrac{1}{2} \leq 1-|\alpha|^n \leq |e^{2\pi it}-\alpha^n| \leq 1+|\alpha|^n \leq 2$$

and that therefore

$$\tfrac{1}{2} \leq M(x-\alpha^n) \leq 2\ .$$

Upon again applying (7) and allowing $n$ to tend to infinity, it follows that also

(b): $$M(x-\alpha) = 1 \quad \text{if} \quad |\alpha| < 1\ .$$

Finally, for $|\alpha| > 1$ ,

$$M(x-\alpha^n) = \exp\left(\int_0^1 \log|e^{2\pi it}-\alpha^n|\,dt\right) = |\alpha|^n \exp\left(\int_0^1 \log|1-\alpha^{-n}e^{2\pi it}|\,dt\right)\ .$$

Here, as $n$ tends to infinity,

$$\lim \log|1-\alpha^{-n}e^{2\pi it}| = 0 \quad \text{uniformly in } t\ ,$$

and hence

$$\lim|\alpha|^{-n}M(x-\alpha^n) = 1 .$$

Therefore, on once more applying the equation (7),

(c): $$M(x-\alpha) = |\alpha| \quad \text{if} \quad |\alpha| > 1 .$$

Combining the three formulae (a), (b), and (c), it follows that

(d): $$M(x-\alpha) = \max(1,|\alpha|) .$$

The product formula (5) for $M(a)$ leads then to the following simple result.

*If* $a(x) = a_0 + a_1x + \ldots + a_mx^m$ *has the zeros* $\alpha_0, \alpha_1, \ldots, \alpha_{m-1}$, *so that*

$$a(x) = a_m(x-\alpha_0)(x-\alpha_1) \ldots (x-\alpha_{m-1}) ,$$

*then*

(8): $$M(a) = |a_m| \prod_{k=0}^{m-1} \max(1,|\alpha_k|) .$$

8. Since

$$|a(e^{2\pi it})| = |a_0+a_1e^{2\pi it}+\ldots+a_me^{2\pi imt}| \le |a_0|+|a_1|+\ldots+|a_m| = L(a) ,$$

$M(a)$ satisfies the inequality

(9): $$M(a) \le L(a) .$$

There is also a similar but less obvious inequality in the opposite direction.

For the coefficient $a_k$ of $a(x)$ can be written as

$$a_k = (-1)^{m-k}a_m\Sigma_{m-k}$$

where $\Sigma_{m-k}$, the $(m-k)$th *elementary symmetric function*, is the sum of all $\binom{m}{m-k} = \binom{m}{k}$ products

$$\alpha_{j_1}\alpha_{j_2} \ldots \alpha_{j_{m-k}} ,$$

where $j_1, j_2, \ldots, j_{m-k}$ are $m-k$ distinct suffixes from the set $0, 1, \ldots, m-1$

Therefore

(10): $$|a_k| \le \binom{m}{k}M(a)$$

because, by (8),

$$|a_m\alpha_{j_1}\alpha_{j_2}\ldots\alpha_{j_{m-k}}| \le M(a) .$$

The formula (10) holds also for $k = 0$ and $k = m$. Since

$$\sum_{k=0}^{m} \binom{m}{k} = 2^m ,$$

we find, on summing over $k = 0, 1, \ldots, m$, that

(11): $$L(a) \le 2^{\partial(a)} M(a) .$$

Finally, from the product equation (5) and the two inequalities (9) and (10), it follows that

$$L(a)L(b) \le 2^{\partial(a)} M(a) . 2^{\partial(b)} M(b) \le 2^{\partial(ab)} M(ab) \le 2^{\partial(ab)} L(ab) ,$$

that is,

(12): $$L(ab) \ge 2^{-\partial(ab)} L(a)L(b) .$$

An analogous inequality

$$H(ab) \ge e^{-\partial(ab)} H(a)H(b)$$

for the height is due to Gelfond (1952).

9. Now let $\xi$ be a real or complex algebraic number, and let

$$A(x|\xi) = A_0 + A_1 x + \ldots + A_M x^M$$

be its minimal polynomial, as defined in §3. Let further

$$a(x) = a_0 + a_1 x + \ldots + a_m x^m , \qquad \text{where } a_m \ne 0 ,$$

be an arbitrary second polynomial with integral coefficients. By definition,

$$A(\xi|\xi) = 0 ,$$

and $A(x|\xi)$ is irreducible. It follows that $a(\xi)$ cannot vanish unless $a(x)$ is divisible by $A(x|\xi)$. Conversely, if $a(\xi)$ does not vanish, then $a(x)$ is not divisible by $A(x|\xi)$, and hence

$$a(\xi^{(j)}) \ne 0 \qquad (j = 0,1,\ldots,M-1) ;$$

here $\xi^{(0)} = \xi, \xi^{(1)}, \ldots, \xi^{(M-1)}$ are again the $M = \partial^0(\xi)$ conjugates of $\xi$. For the remainder of this section, suppose that $a(\xi) \ne 0$.

The resultant

$$R = R(A,a) = A_M^m a(\xi^{(0)}) a(\xi^{(1)}) \ldots a(\xi^{(M-1)})$$

of $A(x|\xi)$ and $a(x)$ is then also distinct from zero. It can be written as the determinant

$$R = \begin{vmatrix} A_0 & A_1 & \cdots & A_M & 0 & \cdots & 0 \\ \vdots & \ddots & & \ddots & & \ddots & \vdots \\ 0 & 0 & \cdots & A_0 & A_1 & \cdots & A_M \\ a_0 & a_1 & \cdots & a_m & 0 & \cdots & 0 \\ \vdots & \ddots & & \ddots & & \ddots & \vdots \\ 0 & 0 & \cdots & a_0 & a_1 & \cdots & a_m \end{vmatrix} \begin{matrix} \left.\vphantom{\begin{matrix}a\\b\\c\end{matrix}}\right\} m \text{ rows} \\ \left.\vphantom{\begin{matrix}a\\b\\c\end{matrix}}\right\} M \text{ rows} \end{matrix}$$

and so is an integer distinct from zero. Hence

$$|R| \geq 1 ,$$

and the trivial estimate

$$|a(\xi^{(j)})| = |a_0 + a_1\xi^{(j)} + \ldots + a_m\xi^{(j)m}| \leq L(a)\max(1, |\xi^{(j)}|)^m$$

shows immediately that

$$\text{(e):} \quad 1 \leq |R| \leq |A_M|^m |a(\xi)| \prod_{j=1}^{M-1} \{L(a)\max(1, |\xi^{(j)}|)^m\} =$$

$$= |a(\xi)| M(A)^m L(a)^{M-1} \max(1, |\xi|)^{-m} .$$

If, however, $\xi$ is non-real, a stronger result can be obtained. For let, say, $\xi^{(1)} = \overline{\xi}$ be the complex conjugate root of $A(x|\xi) = 0$ so that

$$|\xi^{(1)}| = |\xi| , \quad |a(\xi^{(1)})| = |a(\xi)| .$$

We now find that

$$\text{(f):} \quad 1 \leq |R| \leq |A_M|^m |a(\xi)a(\xi^{(1)})| \prod_{j=2}^{M-1} \{L(a)\max(1, |\xi^{(j)}|)^m\} =$$

$$= |a(\xi)|^2 M(A)^m L(a)^{M-2} \max(1, |\xi|)^{-2m} .$$

For any real or complex number $\alpha$ put from now on

$$\sigma(\alpha) = \begin{cases} 1 & \text{if } \alpha \text{ is real,} \\ 2 & \text{if } \alpha \text{ is not real.} \end{cases}$$

The two estimates (e) and (f) can then be combined into the one formula

$$|a(\xi)| \geq \frac{\max(1, |\xi|)^m}{M(A)^{m/\sigma(\xi)} L(a)^{(M/\sigma(\xi))-1}} .$$

Here

$$m = \partial(a) , \quad M = \partial^0(\xi) , \quad M(A) \leq L(A) = L^0(\xi) .$$

We thus arrive at the following theorem which is due to R. Güting (1961).

**THEOREM 1.** *Let* $\xi$ *be an algebraic number, and let* $a(x)$ *be a polynomial with integral coefficients. Then either*

$$a(\xi) = 0$$

*or*

$$|a(\xi)| \geq \frac{\max(1,|\xi|)^{\partial(a)}}{L^0(\xi)^{\partial(a)/\sigma(\xi)} L(a)^{(\partial^0(\xi)/\sigma(\xi))-1}} .$$

**10.** As a first application of this theorem, let the algebraic number $\xi$ be real and irrational, so that $\partial^0(\xi) \geq 2$ and $\sigma(\xi) = 1$ ; let further $a(x)$ be a linear polynomial, say

$$a(x) = p - qx$$

where $p$ and $q > 0$ are integers. From this hypothesis,

$$a(\xi) \neq 0 .$$

Since $L(a) = |p| + q$ , the theorem implies that

$$|p-q\xi| \geq \frac{\max(1,|\xi|)}{L^0(\xi)(|p|+q)^{\partial^0(\xi)-1}}$$

that is,

$$|\xi - \frac{p}{q}| \geq \frac{\max(1,|\xi|)}{L^0(\xi)\left(1 + \frac{|p|}{q}\right)^{\partial^0(\xi)-1}} q^{-\partial^0(\xi)} .$$

Depending on whether

$$|\xi - \frac{p}{q}| < 1 \left(\text{and therefore } 1 + \frac{|p|}{q} < |\xi| + 2\right) \text{ or } |\xi - \frac{p}{q}| \geq 1 ,$$

it follows that

$$|\xi - \frac{p}{q}| \geq \frac{\max(1,|\xi|)}{L^0(\xi)(|\xi|+2)^{\partial^0(\xi)-1}} q^{-\partial^0(\xi)} \quad \text{or} \quad |\xi - \frac{p}{q}| \geq 1 \geq q^{-\partial^0(\xi)} ,$$

respectively. Hence, on putting

$$\gamma(\xi) = \min\left(1, \frac{\max(1,|\xi|)}{L^0(\xi)(|\xi|+2)^{\partial^0(\xi)-1}}\right),$$

where this constant depends only on $\xi$ , we obtain the result that

(13): *If* $\xi$ *is any real irrational algebraic number, then*

$$|\xi - \frac{p}{q}| \geq \gamma(\xi) q^{-\partial^0(\xi)} \quad \textit{for all integers } p \textit{ and } q > 0 .$$

This theorem, but with another estimate for $\gamma(\xi)$ , is due to Liouville (1844). By means of it, he constructed the first example of a transcendental number.

Liouville's theorem evidently implies the following sufficient (but *not* necessary) condition for transcendency.

(14): *Let* $\xi$ *be a real number. If there exist a sequence* $\omega_1, \omega_2, \omega_3, \ldots$ *of real numbers tending to infinity, and a sequence* $\frac{p_1}{q_1}, \frac{p_2}{q_2}, \frac{p_3}{q_3}, \ldots$ *of rational numbers satisfying*

$$q_r \geq 2 , \quad 0 < |\xi - \frac{p_r}{q_r}| \leq q_r^{-\omega_r} \qquad (r = 1,2,3,\ldots) ,$$

*then* $\xi$ *is transcendental.*

In other words, if $\xi$ has *too good rational approximations*, it cannot be algebraic.

Numbers that have the property in (14) are called *Liouville numbers*. It is easy to give an example of a Liouville number. Take, for example, the number

$$\xi = \sum_{n=1}^{\infty} 2^{-n!} .$$

Put

$$p_r = 2^{r!} \sum_{n=1}^{r} 2^{-n!} , \quad q_r = 2^{r!} \qquad (r = 1,2,3,\ldots) .$$

Then

$$0 < \xi - \frac{p_r}{q_r} = \sum_{n=r+1}^{\infty} 2^{-n!} \leq 2^{-(r+1)!}(1+2^{-1}+2^{-2}+\ldots) = 2.2^{-(r+1)r!} ,$$

and therefore, for sufficiently large $r$ ,

$$0 < \xi - \frac{p_r}{q_r} < q_r^{-r} .$$

Thus, by (14), $\xi$ is a Liouville number and so is transcendental.

By generalising this construction, one can obtain a non-countable set of Liouville numbers. On the other hand, one can show that the set of all Liouville numbers has the Lebesgue measure zero.

From its known continued fraction (Perron 1929), *e is not a Liouville number*, and by rather more involved considerations, the same can also be proved for $\pi$ (Mahler 1932). A non-trivial example of a Liouville number is given, for example, by the series

$$\sum_{n=1}^{\infty} \frac{[ne]}{2^n}$$

(Böhmer 1927). Here $[x]$ denotes as usual the integral part of $x$ .

11. Since the time of Liouville, his theorem (13) has successively been improved by Thue (1908), Siegel (1921), Dyson (1947), and Roth (1955). The latter proved the following nearly final and very deep theorem.

(15): *If* $\xi$ *is a real irrational algebraic number, and if* $\tau > 2$ *is a constant, then there exists a constant* $\gamma^*(\xi,\tau) > 0$ *such that*

$$|\xi - \frac{p}{q}| \geq \gamma^*(\xi,\tau)q^{-\tau} \quad \textit{for all integers } p \textit{ and } q > 0 .$$

It is not known whether this theorem remains true when $\tau = 2$ , although this seems rather improbable.

From this theorem follows again a sufficient, but not necessary, condition for transcendency.

(16): *Let* $\xi$ *and* $\tau > 2$ *be real numbers. If there exists an infinite sequence of distinct rational numbers* $\frac{p_1}{q_1}, \frac{p_2}{q_2}, \frac{p_3}{q_3}, \ldots$ *satisfying*

$$0 < \left|\xi - \frac{p_r}{q_r}\right| \leq q_r^{-\tau} \qquad (r = 1,2,3,\ldots) ,$$

*then* $\xi$ *is transcendental.*

This theorem also is useful for the construction of transcendental numbers. With its help one can show (although not quite trivially) that

0.123456789101112...

and infinitely many similar decimal fractions are transcendental (Mahler 1937).

It can be proved that the set of all numbers with the property (16) has the Lebesgue measure zero. This set is, of course, not countable because it contains as a subset the non-countable set of all Liouville numbers.

12. There is another class of transcendental numbers that can be obtained from Theorem 1.

For this purpose let $\xi > 1$ be an algebraic number with the property that none of its powers $\xi, \xi^2, \xi^3, \ldots$ is an integer, and let $a(x)$ run over the binomial polynomials

$$a(x) = x^m - [\xi^m] .$$

Since

$$L(a) = 1 + [\xi^m] ,$$

it follows at once from Theorem 1 that

$$\xi^m - [\xi^m] \geq \frac{\max(1,\xi)^m}{L^0(\xi)^m (1+[\xi^m])^{\partial^0(\xi)-1}} \qquad (m = 1,2,3,\dots) .$$

Hence there exists a constant $C(\xi) > 0$ depending only on $\xi$ such that

(17): $$\xi^m - [\xi^m] \geq C(\xi)^{-m} \qquad (m = 1,2,3,\dots) .$$

This inequality cannot in general be improved, as the following example shows.

Choose

$$\xi = \frac{1+\sqrt{5}}{2} ;$$

it is then easily verified that, for all positive integers $m$ ,

$$[\xi^m] = \begin{cases} \left(\frac{1+\sqrt{5}}{2}\right)^m + \left(\frac{1-\sqrt{5}}{2}\right)^m & \text{if } m \text{ is odd,} \\ \left(\frac{1+\sqrt{5}}{2}\right)^m + \left(\frac{1-\sqrt{5}}{2}\right)^m - 1 & \text{if } m \text{ is even,} \end{cases}$$

and hence that

$$\xi^m - [\xi^m] = \begin{cases} \left(\frac{1+\sqrt{5}}{2}\right)^{-m} & \text{if } m \text{ is odd,} \\ 1 - \left(\frac{1+\sqrt{5}}{2}\right)^{-m} & \text{if } m \text{ is even.} \end{cases}$$

On putting

$$C(\xi) = \frac{1+\sqrt{5}}{2} ,$$

the inequality (17) is then satisfied for all $m$ , and it holds with equality if $m$ is odd.

From (17), one immediately deduces the following sufficient condition for transcendency.

**(18):** *Let* $\xi > 1$ *be such that none of its powers* $\xi, \xi^2, \xi^3, \dots$ *is an integer.*

*If*

$$\liminf_{m\to\infty} \{\xi^m-[\xi^m]\}^{\frac{1}{m}} = 0 ,$$

*then* $\xi$ *is transcendental.*

For this property implies that for every constant $C > 1$ the inequality

$$0 < \xi^m-[\xi^m] \le C^{-m}$$

has infinitely many solutions in positive integers $m$ ; hence, by (17), $\xi$ cannot be algebraic.

The condition (18) is again not necessary for transcendency, as can be proved by means of a counter-example.

There do exist numbers $\xi$ with the property (18), as the following example shows. Define a sequence of positive integers $g_1, g_2, g_3, \ldots$ by

$$g_1 = 1 , \quad g_{r+1} = g_r^{r+1}+1 \qquad (r = 1,2,3,\ldots) ,$$

and put

$$\xi_r = g_r^{\frac{1}{r!}} \qquad (r = 1,2,3,\ldots) ,$$

so that evidently

(19): $$\xi_1 < \xi_2 < \xi_3 < \ldots .$$

We assert that

(20): $$g_r \le 2^{r!-(r-1)!} \qquad (r = 1,2,3,\ldots) .$$

This formula certainly holds if $r = 1$ ; suppose it has already been proved for some suffix $r \ge 1$ . Then

$$g_{r+1} \le 2^{(r+1)\{r!-(r-1)!\}} + 1 = 2^{(r+1)!-r!-(r-1)!} + 1 \le 2^{(r+1)!-r!}$$

because

$$(r+1)! - r! - (r-1)! \ge 0 , \quad (r-1)! \ge 1 ,$$

and so the formula holds also for $r+1$ . Hence (20) is true for all $r$ .

From (20),

$$\xi_r < 2 \qquad (r = 1,2,3,\ldots) .$$

Therefore, by (19), the limit

$$\xi = \lim_{r\to\infty} \xi_r$$

exists and has the property

$$\xi_r < \xi \le 2 \qquad (r = 1,2,3,\ldots) .$$

Since $\xi_2 = \sqrt{2}$ ,

$$\xi_r > \sqrt{2} , \quad g_r > 2^{\frac{1}{2}r!} \quad \text{for} \quad r \ge 3 .$$

It follows that

$$g_{r+1} < g_r^{r+1}\{1+2^{-\frac{1}{2}(r+1)!}\} , \quad \xi_{r+1} < \xi_r\{1+2^{-\frac{1}{2}(r+1)!}\}^{\frac{1}{(r+1)!}}$$

and that therefore

$$0 < \xi_{r+1}-\xi_r < \xi_r\{(1+2^{-\frac{1}{2}(r+1)!})^{\frac{1}{(r+1)!}}-1\} .$$

Now, for $0 < \lambda < 1$ and $x > 0$ , there exists by the mean value theorem of differential calculus a number $\vartheta$ satisfying $0 < \vartheta < 1$ , such that

$$(1+x)^{\lambda} = 1 + \lambda x(1+\vartheta x)^{\lambda-1} \le 1 + \lambda x .$$

Hence, for $r \ge 3$ ,

$$0 < \xi_{r+1} - \xi_r < \xi_r \cdot \frac{2^{-\frac{1}{2}(r+1)!}}{(r+1)!} \le \frac{2^{1-\frac{1}{2}(r+1)!}}{(r+1)!} ,$$

and therefore

$$0 < \xi - \xi_r = \sum_{\rho=r}^{\infty} (\xi_{\rho+1}-\xi_\rho) \le \sum_{\rho=r}^{\infty} \frac{2^{1-\frac{1}{2}(\rho+1)!}}{(\rho+1)!} < \frac{2^{2-\frac{1}{2}(r+1)!}}{(r+1)!} .$$

This further implies that

$$0 < \xi^{r!} - g_r = \xi^{r!} - \xi_r^{r!} = (\xi-\xi_r) \sum_{\rho=0}^{r!-1} \xi^{\rho}\xi_r^{r!-\rho-1} \le (\xi-\xi_r)r!\xi^{r!-1} \le$$

$$\le \frac{2^{2-\frac{1}{2}(r+1)!}}{(r+1)!} r!2^{r!} \le 2^{-\frac{1}{3}(r+1)!} ,$$

as soon as $r$ is sufficiently large.

It follows that both of the relations

$$[\xi^{r!}] = g_r$$

and

$$0 < \xi^{r!}-[\xi^{r!}] < 2^{-\frac{1}{3}(r+1)!} < 1$$

hold as soon as $r$ is sufficiently large. This means, firstly, that no integral power of $\xi$ can be an integer, and secondly, that

$$0 \le \liminf_{m\to\infty} \{\xi^m - [\xi^m]\}^{\frac{1}{m}} \le \liminf_{r\to\infty} \{\xi^{r!} - [\xi^{r!}]\}^{\frac{1}{r!}} \le \lim_{r\to\infty} 2^{-\frac{r+1}{3}} = 0 .$$

Therefore $\xi$ has the property (18) and so is transcendental.

By a small change in the definition of $\xi$, one can construct a non-countable set of numbers with the property (18). It is not known whether any classical constant like $e$ or $\pi$ is of this kind, but this does not seem plausible.

For more results of this kind see the paper Mahler and Szekeres (1968).

**13.** Theorem 1 was concerned with the values, at algebraic points, of polynomials with integral coefficients. Now let $\xi$ be either a transcendental number or an algebraic number of sufficiently high degree, say

$$\partial^0(\xi) > m .$$

Then, if

$$a(x) = a_0 + a_1 x + \dots + a_m x^m$$

is any polynomial not identically zero with integral coefficients and of degree at most $m$, evidently

$$a(\xi) \ne 0 .$$

As will now be shown, the polynomial $a(x)$ can be chosen such that $|a(\xi)|$ is arbitrarily small, except in the one trivial case when $\xi$ is non-real and $m$ is equal to $1$.

There are several ways of constructing such polynomials. The best known one is that due to Dirichlet which is based on his principle, the *Schubfachprinzip* (see p. 92). For our purpose it is more convenient to apply Hermite's method, which depends on estimates for the minima of positive definite quadratic forms. Hermite's original estimates were rather weak, and there are now better ones, in particular those by Minkowski (1910) and Blichfeldt (1917). However, we shall be satisfied with a slightly weaker estimate that is a nearly trivial consequence of Minkowski's theorem on linear forms and which has the advantage of being particularly simple.

**14.** Minkowski's classical theorem on linear forms (Minkowski 1910) is as follows:

**(19):** *Let*

$$l_h(x_1,\dots,x_n) = \sum_{k=1}^{n} l_{hk} x_k \qquad (h = 1,2,\dots,n)$$

*be* $n$ *real linear forms in* $n$ *variables, of determinant*

$$d = \begin{vmatrix} l_{11} & \cdots & l_{1n} \\ \vdots & & \vdots \\ l_{n1} & \cdots & l_{nn} \end{vmatrix} \neq 0 ,$$

*and let* $\lambda_1, \ldots, \lambda_n$ *be* $n$ *positive numbers satisfying*

$$\lambda_1 \ldots \lambda_n \geq |d| .$$

*Then integers* $x_1^0, \ldots, x_n^0$ *not all zero exist such that*

$$|l_h(x_1^0,\ldots,x_n^0)| \leq \lambda_h \qquad (h = 1,2,\ldots,n) .$$

We apply this theorem to a quadratic form

$$F(x_1,\ldots,x_n) = \sum_{h=1}^{n} \sum_{k=1}^{n} F_{hk}x_hx_k$$

with real coefficients

$$F_{hk} = F_{kh} ,$$

which is assumed to be positive definite:

$$F(x_1,\ldots,x_n) > 0 \quad \text{for all real} \quad x_1, \ldots, x_n \quad \text{not all zero.}$$

The discriminant

$$D_F = \begin{vmatrix} F_{11} & \cdots & F_{1n} \\ \vdots & & \vdots \\ F_{n1} & \cdots & F_{nn} \end{vmatrix}$$

of such a positive definite form is always positive. Furthermore, the form can always, in infinitely many ways, be written as a sum

$$F(x_1,\ldots,x_n) = \sum_{h=1}^{n} l_h(x_1,\ldots,x_n)^2$$

of the squares of $n$ real linear forms $l_h(x_1,\ldots,x_n)$ , and then the determinant, $d$ say, of these linear forms is connected with $D_F$ by the equation

$$D_F = d^2 .$$

In Minkowski's theorem (19) choose now

$$\lambda_1 = \ldots = \lambda_n = |d|^{\frac{1}{n}} = D_F^{\frac{1}{2n}} .$$

Then

$$|l_h(x_1^0,\dots,x_n^0)|^2 \le D_F^{\frac{1}{n}} \qquad (h = 1,2,\dots,n)\ ,$$

and hence we obtain the following result.

**(20)**: *If* $F(x_1,\dots,x_n)$ *is a positive definite quadratic form in* $n$ *variables, of discriminant* $D_F$ *, then there exist integers* $x_1^0, \dots, x_n^0$ *not all zero such that*

$$F(x_1^0,\dots,x_n^0) \le nD_F^{1/n}\ .$$

**15.** We apply this theorem to the special quadratic form

$$F(x_1,\dots,x_n) = \sum_{h=1}^{p} (f_{h1}x_1+\dots+f_{hn}x_n)^2 + x_1^2 + \dots + x_n^2$$

where the coefficients $f_{hk}$ may be arbitrary real numbers, and $p$ is an arbitrary positive integer. In the remainder of this section, we use the abbreviation

$$\sum \text{ for } \sum_{h=1}^{p}\ .$$

It is obvious that the quadratic form $F$ is positive definite. Its discriminant

$$D_F = \begin{vmatrix} \sum f_{h1}f_{h1}+1 & \sum f_{h1}f_{h2} & \dots & \sum f_{h1}f_{hn} \\ \sum f_{h2}f_{h1} & \sum f_{h2}f_{h2}+1 & \dots & \sum f_{h2}f_{hn} \\ \vdots & \vdots & & \vdots \\ \sum f_{hn}f_{h1} & \sum f_{hn}f_{h2} & \dots & \sum f_{hn}f_{hn}+1 \end{vmatrix}$$

can be written as a sum of $2^n$ separate determinants by splitting each of its rows

$$\sum f_{hk}f_{h1} \dots \sum f_{hk}f_{hk}+1 \dots \sum f_{hk}f_{hn}$$

into a sum of two rows

$$\sum f_{hk}f_{h1} \dots \sum f_{hk}f_{hk} \dots \sum f_{hk}f_{hn}$$

and

$$0 \dots 1 \dots 0\ .$$

Here, for each suffix $k$ , the term $1$ occurs in the $k$th place.

By this decomposition, $D_F$ takes the form

$$D_F = 1 + \sum_{k=1}^{n} \Delta_k + \sum_{1 \le k_1 < k_2 \le n} \Delta_{k_1 k_2} + \sum_{1 \le k_1 < k_2 < k_3 \le n} \Delta_{k_1 k_2 k_3} + \dots + \Delta_{12\dots n} ,$$

where, for every suffix $\nu$ satisfying $1 \le \nu \le n$, $\Delta_{k_1 k_2 \dots k_\nu}$ denotes the determinant

$$\Delta_{k_1 k_2 \dots k_\nu} = \begin{vmatrix} \sum f_{hk_1} f_{hk_1} & \sum f_{hk_1} f_{hk_2} & \cdots & \sum f_{hk_1} f_{hk_\nu} \\ \sum f_{hk_2} f_{hk_1} & \sum f_{hk_2} f_{hk_2} & \cdots & \sum f_{hk_2} f_{hk_\nu} \\ \vdots & \vdots & & \vdots \\ \sum f_{hk_\nu} f_{hk_1} & \sum f_{hk_\nu} f_{hk_2} & \cdots & \sum f_{hk_\nu} f_{hk_\nu} \end{vmatrix} .$$

In this determinant replace each sum $\sum$ by its full expression. Then $\Delta_{k_1 k_2 \dots k_\nu}$ becomes a sum of $p^\nu$ determinants

$$\sum_{h_1=1}^{p} \sum_{h_2=1}^{p} \cdots \sum_{h_\nu=1}^{p} \begin{vmatrix} f_{h_1k_1} f_{h_1k_1} & f_{h_2k_1} f_{h_2k_2} & \cdots & f_{h_\nu k_1} f_{h_\nu k_\nu} \\ f_{h_1k_2} f_{h_1k_1} & f_{h_2k_2} f_{h_2k_2} & \cdots & f_{h_\nu k_2} f_{h_\nu k_\nu} \\ \vdots & \vdots & & \vdots \\ f_{h_1k_\nu} f_{h_1k_1} & f_{h_2k_\nu} f_{h_2k_2} & \cdots & f_{h_\nu k_\nu} f_{h_\nu k_\nu} \end{vmatrix}$$

and this is the same as

$$\sum_{h_1=1}^{p} \sum_{h_2=1}^{p} \cdots \sum_{h_\nu=1}^{p} \begin{vmatrix} f_{h_1k_1} & f_{h_2k_1} & \cdots & f_{h_\nu k_1} \\ f_{h_1k_2} & f_{h_2k_2} & \cdots & f_{h_\nu k_2} \\ \vdots & \vdots & & \vdots \\ f_{h_1k_\nu} & f_{h_2k_\nu} & \cdots & f_{h_\nu k_\nu} \end{vmatrix} f_{h_1k_1} f_{h_2k_2} \cdots f_{h_\nu k_\nu} .$$

If now two of the suffixes $h_1, h_2, \dots, h_\nu$ are the same, the corresponding determinant vanishes, and if two such suffixes are interchanged, the determinant changes its sign. It follows that

$$\Delta_{k_1k_2\ldots k_\nu} = \sum_{1\le h_1<h_2<\ldots<h_\nu\le p} \begin{vmatrix} f_{h_1k_1} & f_{h_2k_1} & \cdots & f_{h_\nu k_1} \\ f_{h_1k_2} & f_{h_2k_2} & \cdots & f_{h_\nu k_2} \\ \vdots & \vdots & & \vdots \\ f_{h_1k_\nu} & f_{h_2k_\nu} & \cdots & f_{h_\nu k_\nu} \end{vmatrix}^2$$

and so the discriminant $D_F$ can be expressed as the sum

$$(21): \quad D_F = 1 + \sum_{\nu=1}^{n} \sum_{1\le h_1<h_2<\ldots<h_\nu\le p} \sum_{1\le k_1<k_2<\ldots<k_\nu\le n} \begin{vmatrix} f_{h_1k_1} & f_{h_2k_1} & \cdots & f_{h_\nu k_1} \\ f_{h_1k_2} & f_{h_2k_2} & \cdots & f_{h_\nu k_2} \\ \vdots & \vdots & & \vdots \\ f_{h_1k_\nu} & f_{h_2k_\nu} & & f_{h_\nu k_\nu} \end{vmatrix}^2 .$$

Here the inner sums become empty when

$$\nu > \min(n,p) .$$

Thus, for the two smallest values of $p$ we find that

$$(22): \quad D_F = 1 + \sum_{k=1}^{n} f_{1k}^2 \qquad \text{if } p = 1 ,$$

$$(23): \quad D_F = 1 + \sum_{h=1}^{2}\sum_{k=1}^{n} f_{hk}^2 + \sum_{1\le k_1<k_2\le n} \begin{vmatrix} f_{1k_1} & f_{2k_1} \\ f_{1k_2} & f_{2k_2} \end{vmatrix}^2 \qquad \text{if } p = 2 ,$$

respectively.

**16.** As a first application, let $m$ be an arbitrary positive integer and $\xi$ a real number satisfying $\partial^0(\xi) > m$ . Further, let $s$ and $t$ be two parameters satisfying

$$s \ge \max(1,|\xi|)^{-\frac{m}{m+1}} , \quad t = (m+1)(m+2)^{\frac{1}{2(m+1)}} \max(1,|\xi|)^{\frac{m}{m+1}} s ,$$

and hence with the property

$$t \ge (m+1)(m+2)^{\frac{1}{2(m+1)}} .$$

The positive definite quadratic form

$$F(x_0,x_1,\ldots,x_m) = s^{2(m+1)}\left(x_0+x_1\xi+\ldots+x_m\xi^m\right)^2 + x_0^2 + x_1^2 + \ldots + x_m^2$$

is obtained from the general form in §15 on putting

$$n = m+1 \ , \quad p = 1 \ , \quad f_{10} = s^{m+1} \ , \quad f_{11} = s^{m+1}\xi, \ \ldots, \ f_{1m} = s^{m+1}\xi^m \ ,$$

and so, by (22), has the discriminant

$$D_F = 1+s^{2(m+1)}(1+\xi^2+\xi^4+\ldots+\xi^{2m}) \le 1+s^{2(m+1)}(m+1)\max(1,|\xi|)^{2m} \ .$$

Hence, by the lower bound for $s$ ,

$$D_F \le s^{2(m+1)}\max(1,|\xi|)^{2m}+s^{2(m+1)}(m+1)\max(1,|\xi|)^{2m} = s^{2(m+1)}(m+2)\max(1,|\xi|)^{2m} \ .$$

It follows now from (20) that there exist integers $a_0, a_1, \ldots, a_m$ not all zero, with the following property. If $a(x)$ denotes the polynomial

$$a(x) = a_0 + a_1x + \ldots + a_mx^m \ ,$$

then

$$s^{2(m+1)}a(\xi)^2 + a_0^2 + a_1^2 + \ldots + a_m^2 \le (m+1).s^2(m+2)^{\frac{1}{m+1}}\max(1,|\xi|)^{\frac{2m}{m+1}} = \frac{t^2}{m+1} \ .$$

Here

$$a_0^2 + a_1^2 + \ldots + a_m^2 > 0 \quad \text{and hence} \quad a(\xi) \ne 0$$

because $\xi$ is not algebraic of degree at most $m$ . Hence

$$0 < |a(\xi)| < \frac{(m+1)^{\frac{1}{2}}(m+2)^{\frac{1}{2(m+1)}}\max(1,|\xi|)^{\frac{m}{m+1}}}{s^m}$$

and

$$0 < a_0^2 + a_1^2 + \ldots + a_m^2 < \frac{t^2}{m+1} \ .$$

Here the first inequality is equivalent to

$$0 < |a(\xi)| < \frac{(m+1)^{m+\frac{1}{2}}(m+2)^{\frac{1}{2}}\max(1,|\xi|)^m}{t^m}$$

and so implies that

$$0 < |a(\xi)| < \frac{(m+2)^{m+1}\max(1,|\xi|)^m}{t^m} \ .$$

Further, by Cauchy's formula,

$$L(a) = 1.|a_0| + 1.|a_1| + \ldots + 1.|a_m| \le (m+1)^{\frac{1}{2}}(a_0^2+a_1^2+\ldots+a_m^2)^{\frac{1}{2}} \ ,$$

and hence by the second inequality

$$0 < L(a) < t .$$

We have thus proved the following result.

(24): *Let*

$$\sigma(\xi) = 1 , \quad \partial^0(\xi) > m , \quad t \geq (m+1)(m+2)^{\frac{1}{2(m+1)}} .$$

*Then there exists a polynomial* $a(x)$ *with integral coefficients such that*

$$\partial(a) \leq m , \quad 0 < L(a) < t , \quad 0 < |a(\xi)| < \frac{(m+2)^{m+1}\max(1,|\xi|)^m}{t^m} .$$

**17.** As a second application, let $m$ be a positive integer not less than 2 and let $\xi$ be a non-real complex number satisfying $\partial^0(\xi) > m$ . Further, let $s$ and $t$ be two parameters such that

$$s \geq \max(1,|\xi|)^{-\frac{2m}{m+1}} , \quad t = (m+1)(m+2)^{\frac{1}{m+1}}\max(1,|\xi|)^{\frac{2m}{m+1}}s ,$$

and hence

$$t \geq (m+1)(m+2)^{\frac{1}{m+1}} .$$

The expression

$$F(x_0,x_1,\ldots,x_m) = s^{m+1}|x_0+x_1\xi+\ldots+x_m\xi^m|^2 + x_0^2 + x_1^2 + \ldots + x_m^2$$

can be written as the positive definite quadratic form

$$F(x_0,x_1,\ldots,x_m) =$$
$$= s^{m+1}(x_0\lambda_0+x_1\lambda_1+\ldots+x_m\lambda_m)^2 + s^{m+1}(x_0\mu_0+x_1\mu_1+\ldots+x_m\mu_m)^2 + x_0^2 + x_1^2 + \ldots + x_m^2 ,$$

where, for $k = 0, 1, \ldots, m$ , $\lambda_k$ and $\mu_k$ denote the real and the imaginary parts of

$$\xi^k = \lambda_k+i\mu_k ,$$

so that

$$\lambda_k^2+\mu_k^2 = |\xi|^{2k} \quad \text{and} \quad |\lambda_h\mu_k-\lambda_k\mu_h| \leq |\xi|^{h+k} .$$

Evidently $F(x_0,x_1,\ldots,x_m)$ is obtained from the general form in §15 by putting

$$n = m+1 \ , \quad p = 2 \ , \quad f_{10} = s^{\frac{m+1}{2}}\lambda_0, \ \ldots, \ f_{1m} = s^{\frac{m+1}{2}}\lambda_m \ ,$$

$$f_{20} = s^{\frac{m+1}{2}}\mu_0, \ \ldots, \ f_{2m} = s^{\frac{m+1}{2}}\mu_m \ .$$

By (23), this form has the discriminant

$$D_F = 1 + s^{m+1}\sum_{k=0}^{m}(\lambda_k^2+\mu_k^2) + s^{2(m+1)}\sum_{0\le k_1<k_2\le m}(\lambda_{k_1}\mu_{k_2}-\lambda_{k_2}\mu_{k_1})^2 \ .$$

Here

$$\sum_{k=0}^{m}(\lambda_k^2+\mu_k^2) = \sum_{k=0}^{m}|\xi|^{2k} \le (m+1)\max(1,|\xi|)^{2m}$$

and

$$\sum_{0\le k_1<k_2\le m}(\lambda_{k_1}\mu_{k_2}-\lambda_{k_2}\mu_{k_1})^2 < \sum_{k_1=0}^{m}\sum_{k_2=0}^{m}|\xi|^{2(k_1+k_2)} \le (m+1)^2\max(1,|\xi|)^{4m} \ .$$

Hence, by the lower bound for $s$ ,

$$D_F \le 1+s^{m+1}(m+1)\max(1,|\xi|)^{2m}+s^{2(m+1)}(m+1)^2\max(1,|\xi|)^{4m} \le$$

$$\le s^{2(m+1)}\{1+(m+1)+(m+1)^2\}\max(1,|\xi|)^{4m} \le s^{2(m+1)}(m+2)^2\max(1,|\xi|)^{4m} \ .$$

With this upper bound for $D_F$ , we apply again (20). It now follows that there exist integers $a_0, a_1, \ldots, a_m$ not all zero such that the polynomial

$$a(x) = a_0 + a_1x + \ldots + a_mx^m$$

satisfies the inequality

$$s^{m+1}|a(\xi)|^2 + a_0^2 + a_1^2 + \ldots + a_m^2 \le (m+1).s^2(m+2)^{\frac{2}{m+1}}\max(1,|\xi|)^{\frac{4m}{m+1}} \ .$$

Here again

$$a_0^2 + a_1^2 + \ldots + a_m^2 > 0 \quad \text{and hence} \quad a(\xi) \ne 0 \ .$$

Hence

$$0 < |a(\xi)| < \frac{(m+1)^{\frac{1}{2}}(m+2)^{\frac{1}{m+1}}\max(1,|\xi|)^{\frac{2m}{m+1}}}{s^{\frac{m-1}{2}}}$$

and

$$0 < a_0^2 + a_1^2 + \dots + a_m^2 < (m+1)(m+2)^{\frac{2}{m+1}} \max(1,|\xi|)^{\frac{4m}{m+1}} s^2 .$$

As in the real case, the first inequality is equivalent to

$$0 < |a(\xi)| < \frac{(m+1)^{\frac{m}{2}}(m+2)^{\frac{1}{2}}\max(1,|\xi|)^m}{t^{\frac{m-1}{2}}} ,$$

while the second one implies that

$$0 < L(a) \le (m+1)^{\frac{1}{2}}\left(a_0^2+a_1^2+\dots+a_m^2\right)^{\frac{1}{2}} < (m+1)(m+2)^{\frac{1}{m+1}} \max(1,|\xi|)^{\frac{2m}{m+1}} s = t .$$

We have thus the following result.

(25): *Let*

$$\sigma(\xi) = 2 , \quad \partial^0(\xi) > m \ge 2 , \quad t \ge (m+1)(m+2)^{\frac{1}{m+1}} .$$

*Then there exists a polynomial* $a(x)$ *with integral coefficients such that*

$$\partial(a) \le m , \quad 0 < L(a) < t , \quad 0 < |a(\xi)| < \frac{(m+2)^{\frac{m+1}{2}} \max(1,|\xi|)^m}{t^{\frac{m-1}{2}}} .$$

We may finally combine the two results (24) and (25) and formulate the slightly weaker theorem.

THEOREM 2. *Let* $\xi$ *be a real or complex number and* $m$ *an integer such that*

$$\sigma(\xi) \le m < \partial^0(\xi) .$$

*For every value of* $t$ *satisfying*

$$t \ge (m+2)^{1 + \frac{1}{m+1}} ,$$

*there exists a polynomial* $a(x)$ *with integral coefficients satisfying*

$$\partial(a) \le m , \quad 0 < L(a) < t , \quad 0 < |a(\xi)| < \frac{(m+2)^{\frac{m+1}{\sigma(\xi)}} \max(1,|\xi|)^m}{t^{\frac{m+1}{\sigma(\xi)} - 1}} .$$

In this theorem let $t$ increase indefinitely. The polynomial $a(x)$ cannot then remain fixed because the upper bound for $|a(\xi)|$ tends to zero. Thus, by the upper bound $t$ for $L(a)$ , we obtain an *infinite sequence of distinct polynomials* $a(x)$ with integral coefficients for which $\partial(a) \le m$ and

$$0 < |a(\xi)| < \frac{(m+2)^{\frac{m+1}{\sigma(\xi)}} \max(1,|\xi|)^m}{L(a)^{\frac{m+1}{\sigma(\xi)}-1}} .$$

If, on the other hand, $\partial^0(\xi) \leq m$, this proof does not hold, and as soon as $t$ is sufficiently large, $a(x)$ will become a fixed polynomial with the properties

$$a(x) \not\equiv 0 \text{ , but } a(\xi) = 0 .$$

Naturally $a(x)$ will then be divisible by $A(x|\xi)$, the minimal polynomial for $\xi$.

18. The two Theorems 1 and 2 enable us to establish now a condition for transcendency which is both necessary and sufficient. As before, we use the notation

$$\Lambda(a) = 2^{\partial(a)} L(a) .$$

THEOREM 3. *The real or complex number* $\xi$ *is transcendental if and only if there exist (i) an infinite sequence of distinct polynomials*

$$\{a_1(x),\ a_2(x),\ a_3(x),\dots\}$$

*with integral coefficients, and (ii) an infinite sequence of positive numbers*

$$\{\omega_1,\omega_2,\omega_3,\dots\}$$

*tending to infinity, such that*

$$0 < |a_r(\xi)| \leq \Lambda(a_r)^{-\omega_r} \qquad (r = 1,2,3,\dots) .$$

The interest of this theorem lies in the fact that it imposes no restrictions on the degrees $\partial(a_r)$ of the polynomials $a_r(x)$; these may be bounded or unbounded, small or large, as functions of $r$.

Proof. We first show that if $\xi$ has the property formulated in the theorem, then it is a transcendental number. Suppose instead that $\xi$ is algebraic. Then for all $r$, by Theorem 1,

$$\frac{\max(1,|\xi|)^{\partial(a_r)}}{L^0(\xi)^{\frac{\partial(a_r)}{\sigma(\xi)}} L(a_r)^{\frac{\partial^0(\xi)}{\sigma(\xi)}-1}} \leq |a_r(\xi)| \leq \left\{2^{\partial(a_r)} L(a_r)\right\}^{-\omega_r}$$

and hence

$$2^{\partial(a_r)} L(a_r) \leq \max(1,|\xi|)^{-\frac{\partial(a_r)}{\omega_r}} L^0(\xi)^{\frac{\partial(a_r)}{\omega_r\sigma(\xi)}} L(a_r)^{\frac{\partial^0(\xi)-\sigma(\xi)}{\omega_r\sigma(\xi)}} .$$

Here

$$\max(1,|\xi|)^{-\frac{\partial(a_r)}{\omega_r}} \leq 1 .$$

Further, as soon as $r$ and hence also $\omega_r$ are sufficiently large,

$$L^0(\xi)^{\frac{\partial(a_r)}{\omega_r \sigma(\xi)}} \leq 2^{\frac{1}{2}\partial(a_r)}$$

and

$$L(a_r)^{\frac{\partial^0(\xi)-\sigma(\xi)}{\omega_r \sigma(\xi)}} \leq L(a_r)^{\frac{1}{2}} ,$$

because $L(a_r) \geq 1$ .

It follows that for all sufficiently large $r$ ,

$$\Lambda(a_r) \leq \Lambda(a_r)^{\frac{1}{2}} , \quad \Lambda(a_r) \leq 1 .$$

On the other hand, by the hypothesis,

$$\lim_{r\to\infty} \Lambda(a_r) = \infty$$

because the polynomials $a_r(x)$ are all distinct. Hence a contradiction arises.

Secondly, let $\xi$ be a transcendental number. The pair of sequences

$$\{a_r(x)\} \text{ and } \{\omega_r\}$$

can then be constructed in infinitely many ways so as to satisfy the hypothesis of the theorem.

For this purpose let

$$\{m_1, m_2, m_3, \ldots\}$$

be any sequence of positive integers tending to infinity; let $\varepsilon > 0$ be an arbitrarily small positive constant, and let

$$\{t_1, t_2, t_3, \ldots\}$$

be a sequence of positive numbers satisfying

$$t_r \geq (m_r+2)^{1+\varepsilon} \qquad (r = 1,2,3,\ldots)$$

For each suffix $r$ we apply Theorem 2 to $\xi$ , with

$$m = m_r \quad \text{and} \quad t = t_r \ .$$

This we may do as soon as $r$ is sufficiently large, because then

$$t_r \geq (m_r+2)^{1+\varepsilon} \geq (m_r+2)^{1+\frac{1}{m_r+1}} \ .$$

Hence, to every suffix $r \geq r_0$ , there exists a polynomial $a_r(x)$ with integral coefficients, with the property that

$$\partial(a_r) \leq m_r \ , \quad 0 < L(a_r) < t_r \ , \quad 0 < |a_r(\xi)| < \frac{(m_r+2)^{\frac{m_r+1}{\sigma(\xi)}} \max(1,|\xi|)^{m_r}}{t_r^{\frac{m_r+1}{\sigma(\xi)}-1}} \ .$$

Since

$$m_r+2 \leq t_r^{\frac{1}{1+\varepsilon}} \ ,$$

it follows that

$$0 < |a_r(\xi)| < t_r^{\frac{1}{1+\varepsilon}\frac{m_r+1}{\sigma(\xi)} - \left(\frac{m_r+1}{\sigma(\xi)} - 1\right)} \max(1,|\xi|)^{m_r} = t_r^{1 - \frac{\varepsilon(m_r+1)}{(1+\varepsilon)\sigma(\xi)}} \max(1,|\xi|)^{m_r} \ .$$

Here both $m_r$ and $t_r$ tend to infinity, and so

$$t_r \max(1,|\xi|)^{m_r} \leq t_r^{\frac{\varepsilon(m_r+1)}{2(1+\varepsilon)\sigma(\xi)}}$$

as soon as $r$ is sufficiently large. Therefore

$$0 < |a_r(\xi)| < t_r^{-\frac{\varepsilon(m_r+1)}{2(1+\varepsilon)\sigma(\xi)}} \qquad \text{for } r \geq r_1 \ .$$

On the other hand,

$$\Lambda(a_r) = 2^{\partial(a_r)} L(a_r) \leq 2^{m_r} t_r = t_r^{\lambda_r} \ ,$$

where we have put

$$\lambda_r = 1 + \frac{m_r \log 2}{\log t_r} \ .$$

From this definition,

$$\lim_{r\to\infty} \frac{\lambda_r}{m_r} = 0 .$$

Hence, on writing

$$t_r^{-\frac{\varepsilon(m_r+1)}{2(1+\varepsilon)\sigma(\xi)}} = \Lambda(a_r)^{-\omega_r} ,$$

it follows immediately that

$$\lim_{r\to\infty} \omega_r = \infty .$$

This proves that $\xi$ has the required property.

Theorem 3 seems to be of interest on account of its great generality. It can be used for proofs of transcendency of special numbers. It also makes clear one of the difficulties in the theory of transcendental numbers. It will in general not be easy finding sequences of polynomials $a_r(x)$ with integral coefficients for which $|a_r(\xi)|$ tends sufficiently rapidly to zero. Fortunately there is a good deal of latitude in the choice of this sequence, as is clear from the proof of the theorem.

## CHAPTER 2

## CONVERGENT LAURENT SERIES AND FORMAL LAURENT SERIES

**19.** The numbers whose algebraic or transcendental nature one may wish to decide are frequently given as the values $\beta = f(\alpha)$ of analytic functions $f(z)$ of one variable $z$ at *algebraic points* $z = \alpha$ (that is, $\alpha$ is an algebraic number). Here an analytic function may be defined as the set of its elements, that is, of those convergent power series which express the function in some circle.

Let us more generally consider Laurent series

$$f(z) = \sum_{h=\eta}^{\infty} f_h(z-c)^h ,$$

where $f_\eta \neq 0$ , that converge in a domain

$$U_\rho : 0 < |z-c| < \rho .$$

Here $c$ is a complex number, $\rho$ is a positive number, and $\eta$ is an arbitrary integer. If $\eta < 0$ , $f(z)$ has a pole at $z = c$ ; if, however, $\eta \geq 0$ , then $f(z)$ is regular also at the centre $z = c$ of $U_\rho(c)$ . In the latter case, the Laurent series becomes the ordinary Taylor series for $f(z)$ in the disk $|z-c| < \rho$ .

When $\rho = \infty$ , the Laurent series converges in the whole plane except perhaps at $z = c$ , namely, if this point is a pole. If, on the other hand, $\rho$ is finite, then $f(z)$ can possibly be continued outside the domain $U_\rho(c)$ . For the study of the function value $\beta = f(\alpha)$ such a continuation will in general not be necessary, and it will be sufficient to consider $f(z)$ only for values of $z$ in $U_\rho(c)$ .

Some essential properties of the Laurent series defining $f(z)$ are of an algebraic character and have no connection with its convergence. We shall therefore introduce the more general class of formal Laurent series where these properties take a particularly simple form.

**20.** Let $K$ be an arbitrary field of characteristic zero. Of particular importance later will be the fields **Q** of all rational numbers, **A** of all algebraic numbers, **R** of all real numbers, and **C** of all complex numbers. Further, let $z$ be an indeterminate, that is, a quantity transcendental over $K$ . The following notation will be used.

$K[z]$ is the ring of all polynomials in $z$ with coefficients in $K$ .

$K(z)$ is the quotient field of $K[z]$ ; its elements are the rational functions of $z$ with coefficients in $K$ .

$K\langle z-c\rangle$ , where $c$ is in $K$ , is the set of all formal Laurent series

$$f = \sum_{h=\eta}^{\infty} f_h(z-c)^h$$

with coefficients $f_h$ in $K$ , where $\eta$ is some integer depending on $f$ . On putting

$$f_h = 0 \quad \text{for} \quad h < \eta ,$$

the series may be written in the more convenient form

$$f = \sum_{h=-\infty}^{\infty} f_h(z-c)^h .$$

The elements of $K$ play the role of *constants*, while elements of $K[z]$, $K(z)$ , or $K\langle z-c\rangle$, which are not in $K$ , are the *functions*.

The set $K\langle z-c\rangle$ becomes a ring if operations of addition and multiplication are defined as follows. If

$$g = \sum_{h=-\infty}^{\infty} g_h(z-c)^h , \text{ where } g_h = 0 \quad \text{for} \quad h < \rho ,$$

is a second formal Laurent series in $K\langle z-c\rangle$ , put

$$f + g = \sum_{h=-\infty}^{\infty} (f_h+g_h)(z-c)^h$$

and

$$fg = \sum_{h=-\infty}^{\infty} \left(\sum_{k=-\infty}^{\infty} f_k g_{h-k}\right)(z-c)^h .$$

It is then obvious that $f+g$ lies in $K\langle z-c\rangle$ . With regard to $fg$ , the sum

$$\sum_{k=-\infty}^{\infty} f_k g_{h-k}$$

has only finitely many terms distinct from zero, and there are no such terms if $h$ is negative and $|h|$ is sufficiently large. Hence also $fg$ lies in $K\langle z-c\rangle$ . In the special case when $K = C$ and when the series converge in some domain $U_\rho(c)$ , the operations just defined are the same as for convergent series.

It is easily verified that the addition and multiplication in $K\langle z-c\rangle$ are *commutative*, *associative*, and *distributive*.

We identify the element of $K\langle z-c\rangle$ ,

$$f_0 + \sum_{h=1}^{\infty} 0(z-c)^h \quad \text{where} \quad f_0 \in K$$

with the element $f_0$ of $K$ ; then $K$ becomes a subfield of $K\langle z-c\rangle$ . More generally let $a$ be any polynomial in $K[z]$ . This polynomial can be written in the form

$$a = \sum_{h=0}^{m} a_h(z-c)^h$$

where the coefficients $a_h$ are in $K$ . We identify $a$ with the element

$$\sum_{h=0}^{m} a_h(z-c)^h + \sum_{h=m+1}^{\infty} 0(z-c)^h$$

of $K\langle z-c\rangle$ . By this mapping $K[z]$ evidently becomes a subring of $K\langle z-c\rangle$ .

21. It is not difficult to prove that $K\langle z-c\rangle$ is, in fact, a field. For let $f$ be any element of $K\langle z-c\rangle$ which is distinct from $0$ . There is then a suffix $\eta$ such that

$$f_\eta \neq 0 \text{ , but } f_h = 0 \text{ if } h < \eta \text{ .}$$

We call $\eta$ the *order* of $f$ and write

$$\text{ord } f = \eta \text{ .}$$

There exists now a second element

$$f^{-1} = \sum_{h=-\eta}^{\infty} f_h^*(z-c)^h$$

of $K\langle z-c\rangle$ distinct from $0$ , with the property that

$$ff^{-1} = 1 \text{ .}$$

For this relation is equivalent to the infinite system of linear equations

$$f_\eta f_{-\eta}^* = 1 \text{ , } f_\eta f_{-\eta+1}^* + f_{\eta+1} f_{-\eta}^* = 0 \text{ , } f_\eta f_{-\eta+2}^* + f_{\eta+1} f_{-\eta+1}^* + f_{\eta+2} f_{-\eta}^* = 0 \text{ , } \ldots$$

for the coefficients $f_h^*$ of $f^{-1}$ , and these linear equations can be solved step by step for $f_{-\eta}^*, f_{-\eta+1}^*, f_{-\eta+2}^*, \ldots$ , because $f_\eta$ is by hypothesis distinct from zero.

The existence of $f^{-1}$ for every element $f \neq 0$ of $K\langle z-c\rangle$ means that $K\langle z-c\rangle$ is a *field*. As we know already, this field contains the ring $K[z]$ ; hence it also contains the quotient field $K(z)$ of $K[z]$ as a subfield.

22. So far the order of $f$ has been defined only when $f$ is distinct from

0 ; in this exceptional case put

$$\text{ord}\ 0 = \infty\ .$$

With this convention it is easily proved that the order has the following properties:

(1): $$\text{ord}(f+g) \geq \min(\text{ord}\ f,\ \text{ord}\ g)\ ,\quad \text{ord}(fg) = \text{ord}\ f + \text{ord}\ g\ .$$

Thus, in the usual terminology, the order is a non-archimedean valuation of $K\langle z-c\rangle$ .

If in particular $a$ is a polynomial distinct from 0 , it is then obvious that its order and its degree are connected by the inequality

(2): $$\text{ord}\ a \leq \partial(a)\ .$$

**23.** One can also define a formal operation of *differentiation* for the elements of $K\langle z-c\rangle$ . For let

$$f = \sum_{h=\eta}^{\infty} f_h (z-c)^h$$

be a general element of this field. The $n$th derivative of $f$ is then given by

$$\frac{d^n f}{dz^n} = f^{(n)} = \sum_{h=\eta}^{\infty} h(h-1)\ \ldots\ (h-n+1) f_h (z-c)^{h-n} \qquad (n = 1,2,3,\ldots)$$

and so is again in $K\langle z-c\rangle$ . It is not difficult to verify that

(3): $$f^{(n+1)} = \frac{df^{(n)}}{dz}\ ,\quad f^{(n+p)} = \frac{d^p f^{(n)}}{dz^p}\ ,$$

and

(4): $$(f\mp g)' = f' \mp g'\ ,\quad (fg)' = f'g + fg'\ ,\quad \left(\frac{f}{g}\right)' = \frac{f'g-fg'}{g^2}\ .$$

Here $g$ is a second element of $K\langle z-c\rangle$ , which, in the last formula, is assumed to be distinct from 0 .

Evidently

(5): $$\text{ord}\ f' = (\text{ord}\ f) - 1 \quad \text{if} \quad \text{ord}\ f \neq 0\ .$$

If, however, $\text{ord}\ f = 0$ , then $\text{ord}\ f'$ can have any non-negative value. In particular, $f'$ vanishes if and only if $f$ is an element of the constant field $K$ .

**24.** In the special case when $K = C$ is the complex field, denote by $C\{z-c\}$ the set of all series $f$ in $C\langle z-c\rangle$ which converge in some domain

$$U_\rho(c)\ :\ 0 < |z-c| < \rho\ ,$$

where $\rho$ is a positive number that may depend on $f$ ; the case when $\rho = \infty$ is not excluded. As was mentioned already in §20, the operations defined in $C\langle z-c\rangle$ become in $C\{z-c\}$ the addition and multiplication of convergent Laurent series, and the same is of course true for differentiation.

When $f$ is any series in $C\{z-c\}$, we shall always use the convention of denoting by $f(z)$ the analytic function in $U_\rho(c)$ to which this Laurent series converges.

It is obvious that $C\{z-c\}$ is a ring. In fact, it is a field. For let $f \in C\{z-c\}$ be distinct from $0$. Since $z = c$ is at most a pole of $f(z)$, and since $f(z)$ is regular in some domain $U_\rho(c)$, there exists a possibly smaller domain

$$U_\sigma(c) : 0 < |z-c| < \sigma \text{, where } 0 < \sigma \leq \rho \text{,}$$

such that $f(z)$ does not vanish in $U_\sigma(c)$. Hence the reciprocal function $f(z)^{-1}$ is regular in $U_\sigma(c)$ and has at most a pole at $z = c$. It follows that $f(z)^{-1}$ can be developed into a Laurent series convergent in $U_\sigma(c)$, and this Laurent series is exactly the series $f^{-1}$ as defined in $C\langle z-c\rangle$.

25. Let $w_0, w_1, \ldots, w_n$, where $n \geq 0$, be $n+1$ further indeterminates, and let $A(z,w_0,w_1,\ldots,w_n)$ be a polynomial in $z, w_0, w_1, \ldots, w_n$, distinct from $0$, with coefficients in $K$. We shall frequently have to deal with series $f \in K\langle z-c\rangle$ that satisfy a formal algebraic differential equation

$$A(z,f,f',\ldots,f^{(n)}) = 0 \text{,}$$

and in a later chapter we shall study in detail the case when $K = Q$ or $K = A$.

As a special case will be needed soon, we already now prove the following theorem.

(6): *Let $K_0$ be a subfield of $K$, and let $c$ be in $K_0$ and $f$ in $K_0\langle z-c\rangle$. Assume there exists a polynomial $A(z,w_0,w_1,\ldots,w_n)$, not identically zero and with coefficients in $K$, such that*

$$A(z,f,f',\ldots,f^{(n)}) = 0 \text{.}$$

*Then there also exists a polynomial $A_0(z,w_0,w_1,\ldots,w_n)$, not identically zero and with coefficients in $K_0$, such that*

$$A_0(z,f,f',\ldots,f^{(n)}) = 0 \text{.}$$

Proof. The polynomial $A(z,w_0,w_1,\ldots,w_n)$ can be written as a sum

$$A(z,w_0,w_1,\ldots,w_n) = \sum_{\tau=1}^{t} A_\tau z^{\nu_\tau} w_0^{\nu_{\tau 0}} w_1^{\nu_{\tau 1}} \ldots w_n^{\nu_{\tau n}} ,$$

where $t$ is some positive integer, $A_1, \ldots, A_t$ are finitely many elements of $K$ that are all distinct from zero, and the exponents $\nu_\tau, \nu_{\tau 0}, \nu_{\tau 1}, \ldots, \nu_{\tau n}$ are certain non-negative integers. Let $B_1, \ldots, B_t$ be $t$ indeterminates, and put

$$B(z,w_0,w_1,\ldots,w_n) = \sum_{\tau=1}^{t} B_\tau z^{\nu_\tau} w_0^{\nu_{\tau 0}} w_1^{\nu_{\tau 1}} \ldots w_n^{\nu_{\tau n}} .$$

Then

$$B(z,f,f',\ldots,f^{(n)}) = \sum_{\tau=1}^{t} B_\tau g_\tau$$

where the products

$$g_\tau = z^{\nu_\tau} f^{\nu_{\tau 0}} f'^{\nu_{\tau 1}} \ldots f^{(n)^{\nu_{\tau n}}} \qquad (\tau = 1,2,\ldots,t)$$

are certain $t$ elements of $K_0\langle z-c\rangle$, say the series

$$g_\tau = \sum_{h=-\infty}^{\infty} g_{\tau h}(z-c)^h \qquad (\tau = 1,2,\ldots,t) .$$

It follows that

$$B(z,f,f',\ldots,f^{(n)}) = \sum_{h=-\infty}^{\infty} \left( \sum_{\tau=1}^{t} B_\tau g_{\tau h} \right) (z-c)^h .$$

Here the right-hand side is the zero series $0$ if and only if all the homogeneous linear equations

$$\sum_{\tau=1}^{t} B_\tau g_{\tau h} = 0 \qquad (h = 0,\mp 1,\mp 2,\ldots)$$

for $B_1, \ldots, B_t$ are satisfied. Of these infinitely many linear equations at most finitely many can be linearly independent over $K_0$ . Let

$$\sum_{\tau=1}^{t} B_\tau g_{\tau h_\sigma} = 0 \qquad (\sigma = 1,2,\ldots,s)$$

be a maximal system of such linearly independent equations. These equations have coefficients in $K_0$ , and they permit the non-trivial solution

$$B_1 = A_1 \neq 0, \ldots, B_t = A_t \neq 0$$

in $K$ . It follows then that they also possess a solution

$$B_1 = A_{01}, \ldots, B_t = A_{0t}$$

where $A_{01}, \ldots, A_{0t}$ are not all zero and lie in $K_0$ . Now

$$\sum_{\tau=1}^{t} A_{0\tau} g_{\tau h} = 0 \qquad (h = 0, \mp 1, \mp 2, \ldots) ,$$

and so the new polynomial

$$A_0(z, w_0, w_1, \ldots, w_n) = \sum_{\tau=1}^{t} A_{0\tau} z^{\nu_\tau} w_0^{\nu_{\tau 0}} w_1^{\nu_{\tau 1}} \ldots w_n^{\nu_{\tau n}}$$

satisfies the assertion of the theorem.

Let us formulate the special case $n = 0$ of this theorem explicitly; it will soon be needed.

(7): *Let* $K_0$ *be a subfield of* $K$ *, and let* $c$ *be in* $K_0$ *and* $f$ *in* $K_0\langle z-c\rangle$ *. Assume there exists a polynomial* $A(z,w)$ *not identically zero and with coefficients in* $K$ *such that*

$$A(z,f) = 0 .$$

*Then there also exists a polynomial* $A_0(z,w)$ *not identically zero and with coefficients in* $K_0$ *such that*

$$A_0(z,f) = 0 .$$

26. In §1 we introduced the distinction between algebraic numbers and transcendental numbers. There is an analogous distinction for analytic functions and more generally for the formal Laurent series in $K\langle z-c\rangle$ .

An element $f$ of $K\langle z-c\rangle$ is said to be *algebraic*, namely over $K(z)$ , if there exists a polynomial $A(z,w)$ not identically zero with coefficients in $K$ such that

(8): $$A(z,f) = 0 ,$$

and $f$ is called *transcendental* if there is no such polynomial.

Let, in particular, $K = \mathbb{C}$ be the complex field, and let now $f$ be an element of the subfield $\mathbb{C}\{z-c\}$ of $\mathbb{C}\langle z-c\rangle$ . If $f$ is algebraic relative to $\mathbb{C}(z)$ and, say, (8) is one of the algebraic equations satisfied by $f$ , then the corresponding analytic function $f(z)$ satisfies the algebraic equation

$$A\big(z,f(z)\big) = 0$$

identically in the variable $z$ ; we say that $f(z)$ is an *algebraic analytic function*. If, on the other hand, $f$ is transcendental relative to $\mathbb{C}(z)$ , then $f(z)$ is a *transcendental analytic function*.

A number of simple properties are established in function theory that allow one to distinguish between algebraic and transcendental analytic functions. However, these properties have no obvious extensions to the formal Laurent series in the general field $K\langle z-c\rangle$. But it is still possible to establish for the elements of $K\langle z-c\rangle$ a necessary and sufficient condition for transcendency. This condition is of an algebraic nature and is analogous to that in Theorem 3 for the transcendency of complex numbers; also the proof is very similar.

**27.** We commence with analogues of Theorems 1 and 2. The following notation will be used. We denote by $K[z,w]$ the ring of all polynomials in $z$ and $w$ with coefficients in $K$, by $K(z)[w]$ the ring of all polynomials in $w$ that have coefficients in $K(z)$, and by $K\langle z-c\rangle[w]$ the ring of all polynomials in $w$ that have coefficients in $K\langle z-c\rangle$. In all three cases, $\partial_w(a)$ stands for the degree of $a(z,w)$ in $w$; and when $a(z,w)$ is in $K[z,w]$, then $\partial_z(a)$ similarly denotes its degree in $z$. In the latter case we also put

(9): $$\lambda(a) = \partial_z(a) + \partial_w(a) .$$

When $a(z,w)$ is the zero polynomial $0$, we put all three quantities $\partial_z(0)$, $\partial_w(0)$, and $\lambda(0)$ equal to $-\infty$. For the degree of a sum or a product of polynomials, we have in particular

$$\partial_z(a+b) \le \max\big(\partial_z(a),\partial_z(b)\big) , \quad \partial_w(a+b) \le \max\big(\partial_w(a),\partial_w(b)\big) , \quad \lambda(ab) = \lambda(a) + \lambda(b) .$$

Next let $f$ be an arbitrary algebraic element of $K\langle z-c\rangle$. Denote by $\Sigma(f)$ the set of all polynomials

$$A^*(w) = A_0^* + A_1^*w + \dots + A_M^*w^M \qquad \big(A_k^* \in K(z) \text{ for all } k\big)$$

in $K(z)[w]$ for which

$$A_M^* \ne 0 \text{ and } A^*(f) = 0 .$$

There exists in $\Sigma(f)$ at least one polynomial $A^*(w)$ of lowest possible degree, $\partial_w(A^*) = M$ say. Then $f$ is said to be of degree $M$ over $K(z)$, and we write

$$\partial^0(f) = M .$$

Let $A^*(w) \in \Sigma(f)$ be a polynomial of this lowest possible degree $M$. Denote by $d$ the least common denominator of the rational functions $A_0^*, A_1^*, \dots, A_M^*$ that are its coefficients, by $d^*$ the greatest common divisor of their numerators, and put

$$A(z,w) = \frac{d}{d^*}A^*(w) , \quad = A_0 + A_1w + \dots + A_Mw^M \text{ say.}$$

This new polynomial belongs to both $\Sigma(f)$ and $K[z,w]$. It can easily be proved to be irreducible in $K[z,w]$ and to be unique except for a factor in $K$ that does not vanish. We say that $A(z,w)$ so defined is a *minimal polynomial* for $f$. Evidently

$$\partial^0(f) = \partial_w(A) = M .$$

We also put

$$L^0(f) = \partial_z(A) = \max\big(\partial(A_0),\partial(A_1),\dots,\partial(A_M)\big) ,$$

and we call $L^0(f)$ the *length* of $f$.

28. Now let $f \in K\langle z-c\rangle$ be algebraic, and let

$$A(z,w) = A_0 + A_1 w + \dots + A_M w^M$$

be its minimal polynomial. Since $A(z,w)$ vanishes for $w = f$, it allows in $K\langle z-c\rangle[w]$ a factorisation

(10): $$A(z,w) = (w-f)B(w)$$

where $B(w) \in K\langle z-c\rangle[w]$ is a polynomial in $w$ which can be written in the form

$$B(w) = B_0 + B_1 w + \dots + B_{M-1} w^{M-1} .$$

Here $B_0, B_1, \dots, B_{M-1}$ are in $K\langle z-c\rangle$ and in fact have the explicit form

$$B_k = A_{k+1} + A_{k+2} f + \dots + A_M f^{M-k-1} \qquad (k = 0,1,\dots,M-1) .$$

Let us assume from now on that

(11): $$\operatorname{ord} f \geq 0 .$$

Since the polynomials $A_0, A_1, \dots, A_M$ have non-negative orders, it follows that also

(12): $$\operatorname{ord} B_k \geq 0 \qquad (k = 0,1,\dots,M-1) .$$

Now let

$$a(z,w) = a_0 + a_1 w + \dots + a_m w^m$$

be any polynomial in $K[z,w]$ such that

$$a_m \neq 0 , \quad a(z,f) \neq 0 .$$

Its two degrees have the values

$$\partial_z(a) = \max\big(\partial(a_0),\partial(a_1),\dots,\partial(a_m)\big) , \quad \partial_w(a) = m .$$

Since $A(z,w)$ is irreducible and vanishes for $w = f$, while $a(z,w)$ does not, the resultant

$$R(A,a) = \begin{vmatrix} A_0 & A_1 & \cdots & A_M & 0 & \cdots & 0 \\ \vdots & \ddots & & & \ddots & & \vdots \\ 0 & 0 & \cdots & A_0 & A_1 & \cdots & A_M \\ a_0 & a_1 & \cdots & a_m & 0 & \cdots & 0 \\ \vdots & \ddots & & & \ddots & & \vdots \\ 0 & 0 & \cdots & a_0 & a_1 & \cdots & a_m \end{vmatrix} \begin{matrix} \left.\right\} m \text{ rows} \\ \\ \left.\right\} M \text{ rows} \end{matrix}$$

of $A(z,w)$ and $a(z,w)$ with respect to $w$ is not zero. This resultant is a polynomial in $K[z]$, and, from the determinant, its degree satisfies the inequality

$$\partial\big(R(A,a)\big) \leq m\partial_z(A) + M\partial_z(a) .$$

By the earlier relation (2) between order and degree of a polynomial this implies that also

(13): $$\text{ord } R(A,a) \leq m\partial_z(A) + M\partial_z(a) .$$

By a general product formula for resultants the decomposition (10) of $A(z,w)$ implies that

(14): $$R(A,a) = R(w-f,a)\cdot R(B,a) .$$

Further

(15): $$R(w-f,a) = A(z,f) \neq 0 .$$

The resultant $R(B,a)$ can be written as a determinant analogous to that for $R(A,a)$, but with elements that are either series $B_k$ or polynomials $a_k$ or $0$ and so are of non-negative orders, as follows from (12). Hence also

$$\text{ord } R(B,a) \geq 0 .$$

On the other hand, by (14) and (15),

$$\text{ord } R(B,a) = \text{ord } R(A,a) - \text{ord } a(z,f) .$$

It follows then from the estimate (13) that

$$\text{ord } a(z,f) \leq m\partial_z(A) + M\partial_z(a) .$$

This result may be formulated as follows.

(16): *Let $f$ be any algebraic element of $K\langle z-c\rangle$, of order*

$$\text{ord } f \geq 0 .$$

*If $a(z,w)$ is any polynomial in $K[z,w]$, then either*

$$a(z,f) = 0 \ ,$$

*or*

$$a(z,f) \neq 0 \quad \textit{and} \quad \operatorname{ord} a(z,f) \leq L^0(f)\partial_w(a) + \partial^0(f)\partial_z(a) \ .$$

This theorem forms the analogue of Theorem 1 for the field $K\langle z-c\rangle$. It would not be difficult to remove the restriction on $\operatorname{ord} f$ ; this may be left to the reader.

**29.** Let us discuss one special case. We choose for $a(z,w)$ the linear polynomial

$$a(z,w) = w - a_1$$

where $a_1$ is a polynomial in $K[z]$ . In the present case,

$$\partial_z(a) = \partial(a_1) \ , \quad \partial_w(a) = 1 \ .$$

Therefore, if $f \in K\langle z-c\rangle$ is algebraic and $\operatorname{ord} f \geq 0$ , and if further $f$ is not a polynomial in $K[z]$ , then for every $a_1$ in $K[z]$ ,

(17): $$\operatorname{ord}(f-a_1) \leq L^0(f) + \partial^0(f)\partial(a_1) \ .$$

This inequality leads easily to the following sufficient condition for transcendency.

**(18)**: *Let $f \in K\langle z-c\rangle$ not be a polynomial, and let there exist an infinite sequence of polynomials*

$$\{a_1,a_2,a_3,\ldots\}$$

*in $K[z]$ for which*

$$\lim_{r\to\infty} \partial(a_r) = \infty \ , \quad \lim_{r\to\infty} \frac{\operatorname{ord}(f-a_r)}{\partial(a_r)} = \infty \ .$$

*Then $f$ is transcendental.*

For if $r$ is sufficiently large,

$$\operatorname{ord} f = \operatorname{ord}\{(f-a_r)+a_r\} \geq \min\left(\operatorname{ord}(f-a_r),\operatorname{ord} a_r\right) \geq 0 \ ,$$

because $\operatorname{ord} a_r$ cannot be negative. Hence (17) may be applied, giving the assertion.

It is more convenient to express this test for transcendency in a different but equivalent form. Let us say that the series

$$f = \sum_{h=0}^{\infty} f_h(z-c)^h$$

in $K\langle z-c\rangle$ is *lacunary* [resp. *strongly lacunary*] if there exist two sequences of integers

$$\{s_1,s_2,s_3,\dots\} \quad\text{and}\quad \{t_0,t_1,t_2,t_3,\dots\}$$

with the properties

$$0 = t_0 \le s_1 < t_1 \le s_2 < t_2 \le \dots ,$$

$$\lim_{r\to\infty}(t_r-s_r) = \infty \quad \left[\text{resp.}\quad \lim_{r\to\infty}\frac{t_r}{s_r} = \infty\right],$$

$$f_{s_r} \ne 0 , \quad f_{t_r} \ne 0 , \text{ but } f_h = 0 \text{ for } s_r < h < t_r \qquad (r = 1,2,3,\dots) .$$

On applying (18) with

$$a_r = \sum_{h=0}^{s_r} f_h(z-c)^h , \quad f - a_r = \sum_{h=t_r}^{\infty} f_h(z-c)^h ,$$

we immediately obtain the result that

(19): *Every strongly lacunary series is transcendental.*

**30.** So far the degree $\partial^0(f)$ has been defined only when $f \in K\langle z-c\rangle$ is algebraic. When $f$ is transcendental we use the convention of putting

$$\partial^0(f) = \infty$$

where $\infty$ is again considered as larger than any finite integer.

Now let $m$ and $t$ be any two positive integers, and let $f$ be an algebraic or transcendental element of $K\langle z-c\rangle$ such that

$$\operatorname{ord} f \ge 0 , \quad \partial^0(f) > m .$$

By the first of these inequalities, $f$ has a series

$$f = \sum_{h=0}^{\infty} f_h(z-c)^h .$$

Next denote by $a(z,w)$ any polynomial in $K[z,w]$ satisfying

$$\partial_z(a) \le t , \quad \partial_w(a) \le m .$$

The most general polynomial of this kind is

$$a(z,w) = \sum_{h=0}^{t}\sum_{k=0}^{m} a_{hk}z^h w^k ,$$

where the coefficients $a_{hk}$ are in $K$. On substituting $w = f$, $a(z,w)$ becomes an element $a(z,f)$ of $K\langle z-c\rangle$ which can be written as

$$a(z,f) = \sum_{h=0}^{\infty} \alpha_h (z-c)^h .$$

Here the new coefficients $\alpha_h$ are homogeneous linear polynomials in the $(m+1)(t+1)$ parameters $a_{hk}$ with coefficients in $K$. It is evidently possible to choose these $a_{hk}$ in $K$ in such a way that they do not all vanish and so that certain $(m+1)(t+1) - 1$ selected coefficients $\alpha_h$ are zero. Let these be the coefficients for which

$$h = 0,\ 1,\ 2,\ \dots,\ (m+1)(t+1)-2 .$$

It follows that now

$$\operatorname{ord} a(z,f) \geq (m+1)(t+1) - 1 .$$

Furthermore,

$$a(z,f) \neq 0$$

because $\partial^0(f) > m$.

Hence the following theorem holds.

**(20):** *Let $m$ and $t$ be two positive integers, and let $f \in K\langle z-c\rangle$ satisfy the conditions*

$$\operatorname{ord} f \geq 0 , \quad \partial^0(f) > m .$$

*Then there exists a polynomial $a$ in $K[z,w]$ with the properties*

$$\partial_z(a) \leq t , \quad \partial_w(a) \leq m , \quad a(z,f) \neq 0 , \quad \operatorname{ord} a(z,f) \geq mt + m + t .$$

This result is analogous to Theorem 2.

**31.** It is now easy to establish the following necessary and sufficient condition for transcendency.

**THEOREM 4.** *A series $f$ in $K\langle z-c\rangle$ satisfying $\operatorname{ord} f \geq 0$ is transcendental over $K(z)$ if and only if there exist an infinite sequence*

$$\{a_1, a_2, a_3, \dots\}$$

*of non-constant polynomials in $K[z,w]$ and an infinite sequence*

$$\{\omega_1, \omega_2, \omega_3, \dots\}$$

*of positive numbers tending to infinity, such that*

$$a_r(z,f) \neq 0 , \quad \operatorname{ord} a_r(z,f) \geq \omega_r \lambda(a_r) > 0 \qquad (r = 1,2,3,\dots) .$$

Proof. If the condition is satisfied, it implies the transcencency of $f$. For if $f$ is algebraic, then it follows from (16) that

$$\operatorname{ord} a_r(z,f) \le \{L^0(f)+\partial^0(f)\}\lambda(a_r) \qquad (r = 1,2,3,\ldots) .$$

If, on the other hand, $f$ is transcendental, apply (20) with

$$t = m = r \quad \text{for} \quad r = 1,\ 2,\ 3,\ \ldots .$$

We obtain then a sequence of polynomials $a_r \in K[z,w]$ such that

$$a_r(z,f) \ne 0 \ , \quad \lambda(a_r) \le 2r \ , \quad \operatorname{ord} a_r(z,f) > r^2 \qquad (r = 1,2,3,\ldots) ,$$

and this sequence satisfies the assertion.

Theorem 4 is quite analogous to Theorem 3 which deals with real or complex numbers. Although Theorem 4 gives a necessary and sufficient condition for transcendency of formal series, it still leaves many open questions. Thus it does not seem to yield any way of proving the following strengthening of theorem (19):

(21): *Every lacunary series is transcendental.*

We shall base the proof of (21) on the following result due to C.F. Osgood.

(22): *Let $w$ be an algebraic element of degree $n \ge 2$ over $K(z)$. Then $w$ satisfies a linear differential equation of order at most $n - 1$ and with coefficients in $K(z)$.*

Proof of (22). Let

$$(23): \qquad F(w) = w^n + a_{n-1}w^{n-1} + \ldots + a_1 w + a_0 = 0$$

be the minimal polynomial for $w$. Thus, $a_i$ $(0 \le i \le n-1)$ are rational functions of the variable $z$ with coefficients in the field $K$, and the polynomial $F(w)$ is irreducible over $K(z)$. We want to show that $w$ satisfies a linear differential equation

$$(24): \qquad b_{n-1}w^{(n-1)} + b_{n-2}w^{(n-2)} + \ldots + b_1 w' + b_0 w + b = 0 ,$$

where $b_j$ $(0 \le j \le n-1)$ and $b$ are rational functions from $K(z)$, and not all of them are identically zero. Let us differentiate (23) with respect to $z$. We obtain

$$(25): \qquad F_w(w)\cdot w' + G_0(w) = 0 ,$$

where $F_w = \frac{\partial}{\partial w} F$ and $G_0(w)$ is a polynomial with coefficients in $K(z)$. Since $F_w(w)$ is not identically zero and $F(w)$ is irreducible, $F$ and $F_w$ are relatively prime, and the usual identity $a(w)F(w) + b(w)F_w = 1$ shows that when (23) holds,

$F_w(w)$ has a polynomial in $w$ as its reciprocal. After multiplying (25) by this reciprocal, we obtain an equation

(26): $$w' = G_1(w) ,$$

where $G_1$ is a polynomial in $w$ with coefficients in $K(z)$ . Also, we can assume $G_1$ to be of degree at most $n - 1$ since we can reduce it (mod $F$). We proceed by differentiating (26). We obtain

$$w'' = G_2(w) ,$$

where $G_2$ is again a polynomial with coefficients in $K(z)$ and of degree at most $n - 1$ . Namely, we replace $w'$ by (26) and reduce the degree by (23). Continuing this way, we obtain a system

(27): $$w^{(k)} = G_k(w) , \qquad k = 1, 2, \ldots, n-1 ,$$

where $G_k$ $(1 \le k \le n-1)$ are polynomials in $w$ of degrees at most $n - 1$ and with coefficients in $K(z)$ . The system (27) can be written as

(28): $$\begin{array}{rcl} R_{11}w + R_{12}w^2 + \ldots + R_{1,n-1}w^{n-1} & = & w' - R_1 \\ R_{21}w + R_{22}w^2 + \ldots + R_{2,n-1}w^{n-1} & = & w'' - R_2 \\ \vdots & & \\ R_{n-1,1}w + R_{n-1,2}w^2 + \ldots + R_{n-1,n-1}w^{n-1} & = & w^{(n-1)} - R_{n-1} , \end{array}$$

where $R_{ij}$ and $R_i$ $(1 \le i, j \le n-1)$ are elements of $K(z)$ .

Now we shall distinguish two cases. (i) If $\det(R_{ij}) \not\equiv 0$ , then we can solve the system (28) for $w, w^2, \ldots, w^{n-1}$ in terms of the right-hand sides. In particular, we have

(29): $$w = l_0 + l_1w' + l_2w'' + \ldots + l_{n-1}w^{(n-1)} ,$$

where $l_i$ $(0 \le i \le n-1)$ are elements of $K(z)$ , and (29) is just of the form (24). (ii) If $\det(R_{ij}) \equiv 0$ , then the left-hand sides of (28) are linearly dependent over the field $K(z)$ . This means that also the right-hand sides of (28) are linearly dependent. Thus, there are rational functions $l_1, l_2, \ldots, l_{n-1}$ in $K(z)$ , not all identically zero, such that

(30): $$l_1(w'-R_1) + l_2(w''-R_2) + \ldots + l_{n-1}\left(w^{(n-1)}-R_{n-1}\right) = 0 .$$

However, (30) has again the form (24). This completes the proof of (22).

Proof of (21). Suppose now that $w$ has the expansion

(31):
$$w = \sum_{h=0}^{\infty} f_h(z-c)^h \ ,$$

where $c$ and $f_h$ $(h \geq 0)$ are in $K$. If we multiply (24) by the least common denominator of the coefficients $b$ and $b_j$ $(0 \leq j \leq n-1)$, substitute for $w$ its expansion (31) and then compare the coefficients of $(z-c)^h$ $(h = 0,1,2,\ldots)$, we obtain a linear recurrence relation for the coefficients $f_h$, of the form

$$P_m(h)f_{h+m} + P_{m-1}(h)f_{h+m-1} + \ldots + P_1(h)f_{h+1} + P_0(h)f_h = 0 \ , \qquad h \geq h_0 \ ,$$

where $P_m, P_{m-1}, \ldots, P_0$ are polynomials in $h$ over $K$, and $m$ and $h_0$ are fixed integers. Also, $P_m \not\equiv 0$ and hence $P_m(h) \neq 0$ for $h \geq h_0$. It follows that if $m$ consecutive coefficients $f_h, f_{h+1}, \ldots, f_{h+m-1}$ $(h \geq h_0)$ are zero then all after these must be zero. This completes the proof of (21).

32. In the important special case when $K = \mathbb{Q}$ is the field of rational numbers, there is a classical result by Eisenstein (1852) which also can be used to prove the transcencency of power series; in fact, we shall make such an application in the next chapter. Eisenstein's theorem is as follows.

(32): *Let*

$$f = \sum_{h=0}^{\infty} f_h(z-c)^h$$

*be an algebraic element of* $\mathbb{Q}\langle z-c\rangle$. *Then there exists a positive integer* $N$ *such that all the products*

$$N^h f_h \qquad (h = 1,2,3,\ldots)$$

*are integers.*

Thus, in particular, the greatest prime factor $p_h$ of the denominator of the rational number $f_h$ is bounded for all $h$. Conversely, if $p_h$ is not bounded, then $f$ is a transcendental element of $\mathbb{Q}\langle z-c\rangle$.

A proof of Eisenstein's theorem is to be found in Pólya and Szegö (1925).

## CHAPTER 3

## FIRST RESULTS ON THE VALUES OF ANALYTIC FUNCTIONS AT ALGEBRAIC POINTS

**33.** In the notation of the last chapter, let $\sigma = 0$ ; let $K$ be the complex field $\mathbb{C}$ or one of its subfields; and let

$$f = \sum_{h=\eta}^{\infty} f_h z^h$$

be an element of $\mathbb{C}\{z\}$ . This Laurent series converges then to an analytic function $f(z)$ in a domain

$$U_\rho : 0 < |z| < \rho$$

where $\rho$ is positive or equal to $\infty$ . The function $f(z)$ has at most a pole at $z = 0$ and is otherwise regular in $U_\rho$ .

One basic problem in the theory of transcendental numbers may now be formulated as follows.

*Let* $f(z)$ *be given, and let* $S_f$ *be the set of all algebraic points* $\alpha$ *in* $U_\rho$ *for which also the function value* $\beta = f(\alpha)$ *is algebraic. To determine* $S_f$ *, or at least to find properties of this set.*

It is almost unnecessary to stress that this is not an easy problem. We shall in the coming chapters prove a number of theorems about $S_f$ , in particular when $f(z)$ satisfies an algebraic differential equation. As particular cases, these results will contain the transcendency of $e$ and $\pi$ .

In the present chapter we begin by determining $S_f$ for algebraic functions. After this, we shall study some special classes of transcendental functions and disprove in this way certain plausible but false conjectures.

**34.** Let $f$ be an algebraic element of $\mathbb{C}\{z\}$ . Its minimal polynomial $A(z,w)$ is determined only up to a non-zero factor in $\mathbb{C}$ . Two cases must now be distinguished.

(a) The factor can be chosen such that all coefficients of $A(z,w)$ become algebraic.

(b) It is impossible to do so.

In the case (a), $S_f$ *is simply the set of all algebraic points* $z = \alpha$ *in* $U_\rho$. For it is proved in algebra that the field $\mathsf{A}$ of all algebraic numbers is algebraically closed; and we know already that $A(z,w)$ is irreducible over $\mathsf{C}$. If now all the coefficients of $A(z,w)$ are algebraic, and if $z = \alpha$ is an algebraic point in $U_\rho$, then the polynomial $A(\alpha,w)$ does not vanish identically, and the function value $\beta = f(\alpha)$ satisfies the algebraic equation

$$A(\alpha,w) = 0$$

with algebraic coefficients, so that $\beta$ is also algebraic.

A very different result holds in the case (b). Evidently $A(z,w)$ can now be expressed in the form

$$A(z,w) = \sum_{\tau=1}^{t} \gamma_\tau A_\tau(z,w)$$

where $t$ is at least $2$; $\gamma_1, \ldots, \gamma_t$ are $t$ transcendental numbers that are linearly independent over the field $\mathsf{A}$ of algebraic numbers; and $A_1(z,w), \ldots, A_t(z,w)$ are $t$ polynomials in $\mathsf{A}[z,w]$ that are likewise linearly independent over $\mathsf{A}$ and have no common divisor which is a non-constant polynomial. Hence, if both $\alpha$ and $\beta$ are algebraic numbers, the equation

$$A(\alpha,\beta) = 0$$

implies the $t$ equations

$$A_\tau(\alpha,\beta) = 0 \qquad (\tau = 1,2,\ldots,t),$$

and, by the theory of resultants, this system of equations has at most finitely many solutions $(\alpha,\beta)$. Hence in the case (b) the set $S_f$ either is empty or it contains only finitely many points.

We return to the case (a). From what has been proved, we can deduce the following theorem.

(1): *Let*

$$f = \sum_{h=\eta}^{\infty} f_h z^h$$

*be an algebraic series in* $\mathsf{C}\{z\}$ *with algebraic coefficients* $f_h$; *let* $U_\rho$ *be its domain of convergence. Then the set* $S_f$ *consists of all algebraic points in* $U_\rho$.

Proof. Since $f$ is algebraic, it follows from theorem (7) in Chapter 2, applied with $K = \mathsf{C}$ and $K_0 = \mathsf{A}$, that $f$ satisfies an algebraic equation

$$A_0(z,f) = 0$$

where $A_0(z,w)$ is a polynomial in $\mathsf{A}[z,w]$ . The minimal polynomial $A(z,w)$ for $f$ is a divisor of $A_0(z,w)$ and so may also be assumed to have algebraic coefficients. The assertion follows then from what we have already proved.

35. In the remainder of this chapter let the series

$$f = \sum_{h=0}^{\infty} f_h z^h$$

in $\mathsf{C}\{z\}$ be transcendental, have rational coefficients $f_h$ , and have

$$V_\rho : |z| < \rho$$

as its circle of convergence. Here $\rho > 0$ , and the case $\rho = \infty$ is not excluded.

The first investigations of the corresponding set $S_f$ were already made some 80 years ago. According to Stäckel (1895), Strauss in 1886 tried to prove that the analytic function $f(z)$ corresponding to $f$ cannot be rational at all rational points in $V_\rho$ . However, Weierstrass supplied him with a counter-example and also stated that there are transcendental entire functions which assume algebraic values at all algebraic points. Strauss only succeeded in constructing a function that was regular in a *finite* circle and here was algebraic in all algebraic points. The full assertion of Weierstrass was first proved by Stäckel, who established the following more general theorem.

*Let* $\Sigma$ *be a countable set and* $T$ *a dense set in the complex plane. There exists an entire function* $f(z)$ *with rational coefficients* $f_h$ *which assumes only values in* $T$ *at all the points of* $\Sigma$ .

Weierstrass's assertion is obtained in the special case when both $\Sigma$ and $T$ are the set of all algebraic numbers.

In a later paper, Stäckel (1902) found a transcendental function $f(z)$ with rational coefficients $f_h$ , regular in a neighbourhood of the origin, and with the property that *both* $f(z)$ *and its inverse function* assume in this neighbourhood algebraic values at all algebraic points. Further, Faber (1904) constructed an entire transcendental function with rational coefficients and with the property that *this function and all its derivatives* are algebraic at all algebraic points.

In the next sections several of these theorems will be established.

36. We begin with two theorems that can be proved in one and the same way.

(2): *There exists an entire transcendental function*

$$f(z) = \sum_{h=0}^{\infty} f_h z^h$$

*with rational coefficients* $f_h$ *such that* $f(z)$ *and all its derivatives are algebraic at all algebraic points.*

(3): *There exists a transcendental function*

$$g(z) = \sum_{h=0}^{\infty} g_h z^h$$

*with integral coefficients* $g_h$ *which is regular for* $|z| < 1$ *and here with all its derivatives assumes algebraic values at all algebraic points.*

**Proof.** Choose any two power series

$$F = \sum_{h=0}^{\infty} F_h z^h \quad \text{and} \quad G = \sum_{h=0}^{\infty} G_h z^h$$

with positive coefficients of which the first series converges for all $z$, and the second one converges exactly for $|z| < 1$ and has coefficients satisfying

$$\lim_{h\to\infty} G_h = \infty .$$

We found in §4 that the set $\underline{Q}$ of all distinct irreducible primitive normed polynomials of positive degrees and with integral coefficients is countable. With a slight change of notation, let

$$z,\ A_1(z),\ A_2(z),\ A_3(z),\ \ldots$$

be the sequence of all these elements of $\underline{Q}$. Put

$$B_r(z) = \left(A_1(z)A_2(z)\ldots A_r(z)\right)^r \qquad (r = 1,2,3,\ldots) ,$$

denote by

$$d_r = \partial(B_r)$$

the degree of $B_r(z)$, and assume that this polynomial has the explicit form

$$B_r(z) = b_{r0} + b_{r1}z + \ldots + b_{rd_r}z^{d_r}$$

where

$$b_{r0} \neq 0 , \quad b_{rd_r} \neq 0 ,$$

because $B_r(z)$ is not divisible by $z$.

Next let

$$\{a_1,a_2,a_3,\ldots\},\ \{s_1,s_2,s_3,\ldots\},\ \{t_0,t_1,t_2,\ldots\}$$

be three sequences of integers. For the last two sequences we assume that

$$s_r = t_{r-1} + d_r \qquad (r = 1,2,3,\ldots)\ ,$$

$$0 = t_0 \le s_1 < t_1 \le s_2 < t_2 \le s_3 < t_3 \le \ldots\ ,$$

$$\lim_{r\to\infty} (t_r - s_r) = \infty\ .$$

From these conditions, successive polynomials

$$z^{t_{r-1}} B_r(z) = b_{r0} z^{t_{r-1}} + b_{r1} z^{t_{r-1}+1} + \ldots + b_{rd_r} z^{s_r} \qquad (r = 1,2,3,\ldots)$$

involve different powers of $z$ .

Now put

$$f = \sum_{h=0}^{\infty} f_h z^h = \sum_{r=1}^{\infty} \frac{z^{t_{r-1}} B_r(z)}{a_r}$$

where the integers $a_r$ are assumed to increase so rapidly to infinity that

$$|f_h| \le F_h \quad \text{for all} \quad h\ ;$$

such a choice of $a_r$ is evidently possible. Then the convergence of $F$ implies that $f(z)$ is an entire function of $z$ , and $f(z)$ is transcendental because it is not a polynomial. The transcendency of $f(z)$ follows also from theorem (21) in Chapter 2 because the series $f$ is lacunary.

Similarly, put

$$g = \sum_{h=0}^{\infty} g_h z^h = \sum_{r=1}^{\infty} z^{t_{r-1}} B_r(z)$$

where it is now assumed that the integers $t_r$ increase so rapidly that

$$|g_h| \le G_h \quad \text{for} \quad h \ge t_1\ .$$

From the convergence of $G$ it follows that $g(z)$ is regular in the circle $|z| < 1$ . Again $g(z)$ is transcendental because this series is lacunary.

Both series $f$ and $g$ can be differentiated term by term any number of times. For sufficiently large suffix $r$ , the $n$th derivative

$$\frac{d^n}{dz^n}\left\{z^{t_{r-1}} B_r(z)\right\}$$

is divisible both by $z$ and by any other given polynomial $A_k(z)$ in $\underline{\underline{Q}}$ . Hence, when $\alpha$ is any algebraic number, the series for the $n$th derivatives of both $f(z)$ and $g(z)$ at $z = \alpha$ consist of at most finitely many non-zero terms and these terms are polynomials in $\alpha$ with rational coefficients. The assertion follows therefore immediately.

The proof of (3) suggests the following problem which seems difficult.

**PROBLEM A.** *Does there exist a transcendental series*

$$f = \sum_{h=0}^{\infty} f_h z^h$$

*with bounded integral coefficients such that $f(z)$ is algebraic at all algebraic points inside the unit circle $|z| < 1$?*

For such series the unit circle is of course the exact circle of convergence.

I conjecture that this problem has a *negative* answer. In the special case when $f$ is strongly lacunary this can in fact be proved (Mahler 1965).

**37.** We next construct Stäckel's example where both the function and its inverse are algebraic at all algebraic points in a certain neighbourhood of the origin.

Let

$$B_r(z) = \{A_1(z)A_2(z)\dots A_r(z)\}^r = b_{r0} + b_{r1}z + \dots + b_{rd_r}z^{d_r} \qquad (r = 1,2,3,\dots)$$

be the same polynomials as in §36, and let

$$B_r^*(z) = |b_{r0}| + |b_{r1}|z + \dots + |b_{rd_r}|z^{d_r} \qquad (r = 1,2,3,\dots) .$$

Also put

$$u_r = d_1 + d_2 + \dots + d_r + 1 \qquad (r = 1,2,3,\dots)$$

and denote by

$$p_1, p_2, p_3, \dots$$

a sequence of distinct primes which increases so rapidly that the series

$$\sum_{r=1}^{\infty} \frac{(zw)^{u_r}}{p_r} B_r^*(z)B_r^*(w)$$

converges for all positive values of $z$ and $w$. The formula

$$\Phi(z,w) = z + w + \sum_{r=1}^{\infty} \frac{(zw)^{u_r}}{p_r} B_r(z)B_r(w)$$

defines then an entire function of $z$ and $w$, and this function has the properties

$$\Phi(z,w) = \Phi(w,z) \;;\quad \Phi(0,0) = 0 \;;\quad \left.\frac{\partial\Phi(z,w)}{\partial w}\right|_{z=w=0} = 1 .$$

The last equation holds because $\Phi(z,w)$ allows for all $z$ and $w$ a convergent development of the form

$$\Phi(z,w) = z + w + \sum_{h=2}^{\infty} \sum_{k=2}^{\infty} \varphi_{hk} z^h w^k$$

with certain rational coefficients $\varphi_{hk}$.

By the theorem on implicit functions there exists then a neighbourhood

$$U_\rho : |z| < \rho$$

of $z = 0$ in which the equation

$$\Phi(z,w) = 0$$

can be solved in the form of a convergent power series

$$w = f = -z + \sum_{h=2}^{\infty} f_h z^h .$$

Here $f(z)$ is regular in $U_\rho$, and the equation

$$w = f(z)$$

can be solved for $z$ by a relation of exactly the same form,

$$z = f(w) ,$$

because $\Phi(z,w)$ is symmetrical in $z$ and $w$.

We can differentiate both the equation $\Phi(z,w) = 0$ and the relations $w = f(z)$ and $z = f(w)$ any number of times term by term. Upon then substituting for $z$ or $w$ any algebraic number $\alpha$ in $U_\rho$, the differentiated infinite series for $\Phi$ reduces to a sum of only finitely many non-zero terms. Therefore all the derivatives

$$f^{(n)}(\alpha) \qquad (n = 0,1,2,3,\ldots)$$

satisfy algebraic equations with algebraic coefficients and hence have algebraic values. The same result trivially holds for the inverse function $z = f(w)$.

The result so obtained would be without interest if the function $w = f(z)$ and hence also its inverse function should turn out to be algebraic functions. It is exactly at this point where the actual difficulty of Stäckel's theorem lies. The function $w = f(z)$ satisfying the transcendental equation $\Phi(z,w) = 0$ is, at the outset, defined only in $U_\rho$, where it is given by its convergent power series. It may be that it can be continued outside this circle and that it then no longer remains single-valued. However, a full investigation of its function-theoretical properties does not seem to be easy, and one has to look for other methods in order to decide whether $w = f(z)$ is algebraic or transcendental.

Stäckel did so by means of an algebraic method. He selected the exponents $u_r$ not as we have done, but in such a way that they tend very rapidly to infinity. The transcendency of $w = f(z)$ could then be established by means of Hilbert's irreducibility theorem for polynomials in several variables, but this proof was not very simple.

We shall instead base our proof of transcendency on Eisenstein's theorem (32) in Chapter 2 and on simple arithmetic considerations.

The proof is indirect; that is, we assume that the function

$$w = f(z) = -z + \sum_{h=2}^{\infty} f_h z^h$$

is algebraic. By Eisenstein's theorem, this implies that there exists a positive integer $N$ such that none of the denominators of the coefficients $f_h$ can be divisible by a prime greater than $N$.

On the primes $p_r$ of the sequence $\{p_1, p_2, p_3, \ldots\}$ we now impose the additional conditions

$$p_r > \max\left(|b_{r0}|, |b_{r1}|, \ldots, |b_{rd_r}|\right) \qquad (r = 1,2,3,\ldots),$$

as is evidently permitted. On replacing $w = f(z)$ in $\Phi(z, w) = 0$ by its power series, we obtain the equation

$$\sum_{h=2}^{\infty} f_h z^h + \sum_{r=1}^{\infty} \frac{z^{u_r}}{p_r} \left(\sum_{j=0}^{d_r} b_{rj} z^j\right) \sum_{k=0}^{d_r} b_{rk}\left(-z + \sum_{h=2}^{\infty} f_h z^h\right)^{k+u_r} = 0$$

which holds identically in $z$.

Now let $s$ be an integer so large that

$$p_s > N .$$

The coefficient of $z^{2u_s}$ on the left-hand side of this identity evidently has the form

$$f_{2u_s} + \frac{(-1)^{u_s}}{p_s} b_{s0}^2 + \sum_{r=1}^{s-1} \frac{P_r}{p_r}, \quad = \Sigma \quad \text{say},$$

where $P_r$, a polynomial in finitely many coefficients $f_h$ with integral coefficients, is a rational number the denominator of which has no prime factor greater than $N$. By hypothesis the coefficient $f_{2u_s}$ is a rational number of the same kind, and so the denominator of the sum

$$f_{2u_s} + \sum_{r=1}^{s-1} \frac{P_r}{p_r}$$

cannot be divisible by $p_s$ because

$$0 < p_1 < p_2 < \dots < p_{s-1} < p_s \,.$$

On the other hand, $b_{s0}$ is an integer not zero and not divisible by $p_s$. Hence the terms of $\Sigma$ do not cancel each other, so that $\Sigma$ is distinct from zero, and a contradiction arises. Thus we obtain the following theorem.

(4): *There exists a transcendental series*

$$f = -z + \sum_{h=2}^{\infty} f_h z^h \,,$$

*with rational coefficients, which converges in a neighbourhood of the origin and has the property that both $f(z)$ and its inverse function, as well as all their derivatives, are algebraic at all algebraic points in this neighbourhood.*

This result suggests the following question.

**PROBLEM B.** *Does there exist an entire transcendental function*

$$f(z) = \sum_{h=0}^{\infty} f_h z^h$$

*with rational coefficients $f_h$ such that both $f(z)$ and its inverse function are algebraic in all algebraic points?*

It even seems to be unknown whether there exists an entire function of this kind where the coefficients $f_h$ may be arbitrary *complex* numbers.

**38.** A convergent power series with *real* coefficients always assumes, at complex conjugate points $\zeta$ and $\overline{\zeta}$, values that are likewise complex conjugate. In particular, if $\zeta$ is a zero, the complex conjugate point $\overline{\zeta}$ also is a zero.

Conversely, let

$$Z = \{\zeta_1, \zeta_2, \zeta_3, \dots\}$$

be a finite or infinite sequence of complex numbers with the following two properties.

(5): If $\zeta$ is an element of $Z$ and occurs $k$ times in the sequence $Z$, then the complex number $\overline{\zeta}$ likewise belongs to $Z$ and occurs $k$ times.

(6): If the sequence $Z$ is infinite, then

$$\lim_{r\to\infty} |\zeta_r| = \infty .$$

It follows from Weierstrass's classical theorem on entire functions that under these assumptions there exists an entire function

$$F(z) = \sum_{h=0}^{\infty} F_h z^h$$

with real coefficients $F_h$ for which $Z$ is the exact sequence of its zeros; in particular, elements that occur $k$ times in $Z$ are zeros of $F(z)$ of order $k$.

Hurwitz (1891) deduced from Weierstrass's theorem the following surprising result.

(7): *If the sequence* $Z$ *has the properties* (5) *and* (6), *then there exists an entire function*

$$f(z) = \sum_{h=0}^{\infty} f_h z^h$$

*with rational coefficients such that* $Z$ *is the exact sequence of zeros of* $f(z)$ .

Proof. Without loss of generality, the origin $z = 0$ is not an element of $Z$, and the function $F(z)$ as given by Weierstrass's theorem has the constant term

$$F_0 = 1 .$$

For the moment denote by

$$G = \sum_{h=1}^{\infty} G_h z^h$$

an arbitrary formal power series with constant term 0, and by

$$e^G = \sum_{k=0}^{\infty} \frac{G^k}{k!} = \sum_{h=0}^{\infty} H_h z^h$$

its formal exponential series. The coefficients $H_h$ are then

$$H_0 = 1 , \quad H_1 = G_1 , \quad H_2 = G_2 + \tfrac{1}{2}G_1^2 , \quad H_3 = G_3 + G_1G_2 + \tfrac{1}{6}G_1^3 ,$$

and so on, and generally $H_h$ can be written in the form

$$H_h = G_h + p_h(G_1, G_2, \ldots, G_{h-1}) \qquad (h = 1,2,3,\ldots)$$

where $p_h$ is a certain polynomial in $G_1, G_2, \ldots, G_{h-1}$ with real coefficients.

Now put

$$f = \sum_{h=0}^{\infty} f_h z^h = Fe^G .$$

Since $F_0 = 1$ , also $f_0 = 1$ , and the remaining coefficients $f_h$ are given recursively by

$$f_h = F_h + F_{h-1}H_1 + \ldots + F_1 H_{h-1} + H_h \qquad (h = 1,2,3,\ldots) .$$

They may therefore be written in the form

(8): $$f_h = G_h + q_h(F_1, F_2, \ldots, F_h, G_1, G_2, \ldots, G_{h-1}) \qquad (h = 1,2,3,\ldots) ,$$

where $q_h$ is some polynomial in $F_1, F_2, \ldots, F_h, G_1, G_2, \ldots, G_{h-1}$ with real coefficients.

By assumption $F(z)$ is an entire function. The same will therefore be true for $f(z)$ if the series $G$ converges for all $z$ , and to ensure this we impose the restriction that the coefficients of $G$ satisfy the inequalities

$$0 \leq G_h \leq \frac{1}{h!} \qquad (h = 1,2,3,\ldots) .$$

Since the rational numbers are dense on the real axis, it is moreover possible to choose the coefficients $G_h$ in their smaller and smaller intervals so that all the coefficients $f_h$ become rational. Evidently both $G(z)$ and $e^{G(z)}$ are now entire functions, and $e^{G(z)}$ has no zeros in the complex plane, whence the assertion.

Hurwitz's theorem shows that no general arithmetic statement about the zeros of an entire function with rational coefficients can be made. Thus such a function may vanish at a certain algebraic point $\zeta$ , and if then $\zeta'$ is algebraically conjugate, but not complex conjugate, to $\zeta$ , the function need not vanish at $\zeta'$ .

**39.** Hurwitz's theorem has an analogue for functions defined in a finite region.

**(9):** *Let*

$$F(z) = \sum_{h=0}^{\infty} F_h z^h ,$$

*where* $F_0 = 1$ *, be a power series with real coefficients that converges in the circle*

$$U_\rho : |z| < \rho$$

*where* $0 < \rho < 1$ *. There exists a power series*

$$f(z) = \sum_{h=0}^{\infty} f_h z^h$$

*with integral coefficients which likewise converges in* $U_\rho$ *, and which has the same zeros as* $F(z)$ *has.*

Proof. This time we choose the coefficients $G_h$ so that

$$0 \le G_h \le 1 \qquad (h = 1,2,3,\ldots)$$

and so that simultaneously all coefficients $f_h$ become integers. Then both $G(z)$ and $e^{G(z)}$ are regular in $|z| < 1$ , and $e^{G(z)}$ does not vanish, whence the assertion.

In the result just proved the Taylor coefficients $f_h$ will not in general be *bounded*. If, however, suitable restrictions are imposed on the zeros of $f(z)$ , then results analogous to (9) with bounded integral $f_h$ can be established. I refer, without proof, to a result by A.O. Gelfond (1965):

(10): *Let* $c$ *be a constant satisfying* $0 < c < 1$ *, and let* $\{\zeta_r\}$ *be an infinite sequence of real or complex numbers such that*

$$0 < |\zeta_r| < 1 , \quad |\zeta_1\zeta_2\ldots\zeta_r| \ge c \qquad (r = 1,2,3,\ldots) .$$

*Then there exists a power series*

$$f = \sum_{h=0}^{\infty} f_h z^h$$

*with bounded integral coefficients* $f_h$ *such that*

$$f_0 \ne 0 , \quad f(\zeta_r) = 0 \quad \textit{for all} \quad r .$$

**40.** We next prove a result due to C.G. Lekkerkerker (1949). This result also is concerned with power series that have bounded integral coefficients.

(11): *Let* $(\alpha_1,\beta_1), (\alpha_2,\beta_2), \ldots, (\alpha_n,\beta_n)$ *be finitely many pairs of complex numbers which satisfy the following conditions.*

*(a): The numbers* $\alpha_1, \alpha_2, \ldots, \alpha_n$ *are all distinct.*

*(b):* $$0 < |\alpha_k| < 1 \qquad (k = 1,2,\ldots,n) .$$

*(c): Whenever* $\alpha_k$ *is real, also* $\beta_k$ *is real.*

*(d): When* $\alpha_k$ *is not real, there is a suffix* $l$ *such that* $\alpha_l$ *and* $\beta_l$ *are complex conjugate to* $\alpha_k$ *and* $\beta_k$ *, respectively.*

*Then there exists a power series*

$$f = \sum_{h=0}^{\infty} f_h z^h$$

*with bounded integral coefficients such that*

$$f(\alpha_k) = \beta_k \qquad (k = 1,2,\ldots,n) .$$

**Proof.** Put

$$a = (z-\alpha_1)(z-\alpha_2) \ldots (z-\alpha_n) = a_0 + a_1 z + \ldots + a_n z^n ,$$

so that $a$ is a polynomial with real coefficients and

$$a_0 = (-1)^n \alpha_1 \alpha_2 \ldots \alpha_n \neq 0 .$$

Next, there is a polynomial

$$b = b_0 + b_1 z + \ldots + b_{n-1} z^{n-1}$$

of degree at most $n - 1$ such that

$$b(\alpha_k) = \beta_k \qquad (k = 1,2,\ldots,n) .$$

This polynomial also has real coefficients because, by Lagrange's formula,

$$b = \sum_{k=1}^{n} \frac{a(z)\beta_k}{a'(\alpha_k)(z-\alpha_k)} .$$

Denote now by

$$g = \sum_{h=0}^{\infty} g_h z^h$$

a formal power series still to be chosen, and put

$$f = b + ag = \sum_{h=0}^{\infty} f_h z^h .$$

This relation between power series is equivalent to the equations

$$f_0 = b_0 + a_0 g_0 ,$$
$$f_1 = b_1 + a_1 g_0 + a_0 g_1 ,$$
$$\vdots$$
$$f_{n-1} = b_{n-1} + a_{n-1} g_0 + a_{n-2} g_1 + \ldots + a_0 g_{n-1} ,$$

and

$$f_h = a_n g_{h-n} + a_{n-1} g_{h-n+1} + \dots + a_0 g_h \qquad (h = n, n+1, n+2, \dots) .$$

Since $a_0 \neq 0$ , it is possible to choose successive real values for $g_0, g_1, g_2, \dots$ in such a way that

$$-\frac{1}{2|a_0|} \leq g_h \leq +\frac{1}{2|a_0|} \qquad (h = 0,1,2,\dots)$$

and that at the same time all coefficients $f_0, f_1, f_2, \dots$ become integers. Then evidently

$$|f_h| \leq (|a_0|+|a_1|+\dots+|a_n|) \max_{k=0,1,2,\dots} |g_k| \leq \frac{L(a)}{2|a_0|} \qquad (h = n, n+1, n+2, \dots) ,$$

and so $f$ has *bounded* coefficients; and the same is of course true for $g$ . It follows that both series $f$ and $g$ are convergent in the unit circle $|z| < 1$ . Further,

$$f(\alpha_k) = b(\alpha_k) + a(\alpha_k)g(\alpha_k) = \beta_k \qquad (k = 1,2,3,\dots) .$$

Hence the series $f$ as constructed has the required properties.

By way of example, let $n = 2$ , and let $\alpha_1$ and $\alpha_2$ be conjugate algebraic numbers that are not complex conjugates, hence are real. We may then choose for $\beta_1$ a real algebraic number and for $\beta_2$ a real transcendental number. Thus the behaviour of a power series with bounded integral coefficients at one algebraic point does not necessarily say anything about its behaviour at a conjugate algebraic point!

It would be of some interest to extend the theorem just proved to infinite sets of pairs $(\alpha_k, \beta_k)$ .

**41.** Denote for the moment by $T_\rho$ the set of all power series

$$f = \sum_{h=0}^{\infty} f_h z^h$$

with rational coefficients that converge in

$$U_\rho : |z| < \rho .$$

Further, let $S$ be any finite or infinite set of algebraic points $z = \alpha$ in $U_\rho$ . The following question arises.

**PROBLEM C.** *Does there exist for every choice of* $S$ *a series* $f$ *in* $T_\rho$ *which is algebraic in all points of* $S$ *and in no other algebraic points, thus for which* $S_f = S$ *?*

The answer for general sets $S$ seems to be unknown. There is, however, one special class of sets $S$ for which a complete answer can be given.

Let us say that the set $S$ is *closed in* $U_\rho$ if it has the following property: if $\alpha \in U_\rho$ is algebraic, and if $\alpha'$ is any algebraic conjugate of $\alpha$ that also lies in $U_\rho$, then $\alpha \in S$ implies that also $\alpha' \in S$.

With this notation, the following theorem can be proved.

(12): *If the set* $S$ *is closed in* $U_\rho$, *then there exists a series* $f$ *in* $T_\rho$ *such that* $S_f = S$.

In fact, more can be proved. First let $0 < \rho \leq 1$. Then one may choose for $f$ even a *strongly lacunary series with integral coefficients*; and, conversely, for such a series the set $S_f$ is always closed in its circle of convergence $U_\rho$. When $\rho > 1$, $f$ may still be chosen as a strongly lacunary series, but its coefficients will no longer all be integers.

For a proof of this theorem see my paper (Mahler, 1965).

# CHAPTER 4

## LINEAR DIFFERENTIAL EQUATIONS: THE LEMMAS OF SHIDLOVSKI

**42.** As was mentioned earlier, Liouville published the first explicit examples of transcendental numbers in 1844. Not quite thirty years later, the technique of proofs of transcendency was put on a firm basis by Hermite (1873) and Lindemann (1882) in their work on $e$ and $\pi$, respectively. After another fifty years, the next great progress was made by Siegel (1929; see also his book of 1949). He established an entirely new and very powerful method for proofs of transcendency and applied this method to the study of analytic functions satisfying linear differential equations of the first and second orders with rational functions as coefficients. The best known of his results deals with Bessel functions. Let, for instance,

$$w = J_0(z) = \sum_{h=0}^{\infty} \frac{\left(-\frac{z^2}{4}\right)^h}{h!h!}$$

be the Bessel function of order zero which satisfies the differential equation

$$zw'' + w' + zw = 0 .$$

If $\alpha \neq 0$ is an algebraic number, then not only are $J_0(\alpha)$ and $J_0'(\alpha)$ both transcendental, but these two function values are not even connected by any algebraic equation with algebraic coefficients.

Since 1954, in a series of important papers, Shidlovski (see, in particular, his papers 1959a, 1959b, and 1962) simplified Siegel's method and extended it to the most general system of inhomogeneous linear differential equations

$$w_h' = q_{h0} + \sum_{k=1}^{m} q_{hk}w_k \qquad (h = 1,2,\ldots,m)$$

with rational functions $q_{h0}$ and $q_{hk}$ as coefficients. The algebraic basis of his method will be discussed in detail in the present chapter. In the next two chapters his method will then be combined with the ideas of Siegel and will lead both to general theorems and to results on special functions.

**43.** In the present chapter we are concerned with systems of *homogeneous* linear differential equations

$$w'_h = \sum_{k=1}^{m} q_{hk} w_k \qquad (h = 1,2,\ldots,m)$$

where the coefficients $q_{hk}$ are rational functions of $z$. But instead of considering the analytic functions that satisfy such a system, we shall study solutions in *formal power series*. This approach has the advantage of stressing the algebraic side of the problem and of disregarding less essential properties connected with analyticity.

Two main results will be proved in this chapter. The first deals with a certain reduction of the given system of differential equations. The second establishes sufficient conditions under which a certain determinant is distinct from zero. These two results represent perhaps Shidlovski's most important contributions to the theory. As will be seen, the proofs are highly ingenious.

The same notation as in §§20-23 in Chapter 2 will be used. In particular, $K$, $K[z]$, $K(z)$, $K\langle z-c\rangle$, $\partial(a)$, and $\operatorname{ord} f$ have the same meaning as before. The elements of $K$, $K[z]$, $K(z)$, and $K\langle z-c\rangle$ will for simplicity be called *constants*, *polynomials*, *rational functions*, and *series*, respectively. We shall also occasionally speak of constant vectors, polynomial vectors, and rational vectors when the components of these vectors lie in $K$, $K[z]$, and $K(z)$, respectively, and we use similar terms for matrices.

A rational function $r$ can be written as the quotient $r = a/b$ of two polynomials $a$ and $b \neq 0$ that are relatively prime; here $a$ and $b$ are unique except for a common constant factor. The maximum

$$\nabla(r) = \max\big(\partial(a),\partial(b)\big)$$

depends therefore only on $r$ and not on the special choice of $a$ and $b$. $\nabla(r)$ is called the *degree* of the rational function $r$. This name is justified because when $r$ is a polynomial, then evidently $\nabla(r) = \partial(r)$, except only when $r = 0$.

Next denote by $K_0$ any field satisfying

$$K \subseteq K_0 \subseteq K\langle z-c\rangle ,$$

and by $f_1, \ldots, f_n$ finitely many series in $K\langle z-c\rangle$. If there exist elements $a_1, \ldots, a_n$ in $K_0$ not all zero such that

$$a_1 f_1 + \ldots + a_n f_n = 0 ,$$

then, as usual, $f_1, \ldots, f_n$ are called *linearly dependent over* $K_0$, and they are said to be *linearly independent over* $K_0$ if no such elements $a_1, \ldots, a_n$ exist.

More generally, finitely many column vectors

$$\mathbf{f}_1 = \begin{bmatrix} f_{11} \\ \vdots \\ f_{m1} \end{bmatrix}, \ldots, \mathbf{f}_n = \begin{bmatrix} f_{1n} \\ \vdots \\ f_{mn} \end{bmatrix}$$

with components in $K\langle z-c\rangle$ are said to be linearly dependent over $K_0$ if there are elements $a_1, \ldots, a_n$ in $K_0$ not all zero such that

$$a_1\mathbf{f}_1 + \ldots + a_n\mathbf{f}_n = 0 = \begin{bmatrix} 0 \\ \vdots \\ 0 \end{bmatrix},$$

and they are called linearly independent over $K_0$ in the opposite case. The same terminology is applied to row vectors. When this notation is used in the present chapter, $K_0$ will usually be one of the three fields $K$, $K(z)$, or $K\langle z-c\rangle$.

**44.** Since it will later be needed, we begin with a simple lemma.

(1): *Let* $\psi_1, \ldots, \psi_u$ *be finitely many series in* $K\langle z-c\rangle$, *and let* $n \geq 0$ *be a given integer. Then there exists a positive integer* $\Gamma$ *with the following property: if* $a_1, \ldots, a_u$ *are any* $u$ *polynomials of degrees at most* $n$ *and satisfying*

$$a_1\psi_1 + \ldots + a_u\psi_u \neq 0,$$

*then*

$$\operatorname{ord}(a_1\psi_1+\ldots+a_u\psi_u) \leq \Gamma .$$

Proof. Put $N = (n+1)u$, and denote by $\chi_1, \ldots, \chi_N$ the $N$ series

$$\psi_1, z\psi_1, \ldots, z^n\psi_1, \ldots, \psi_u, z\psi_u, \ldots, z^n\psi_u$$

arranged in this order. If the assertion is false, then $N+1$ systems of $N$ constants

$$b_{\nu 1}, \ldots, b_{\nu N} \qquad (\nu = 0,1,\ldots,N)$$

exist such that all the sums

$$\Psi_\nu = b_{\nu 1}\chi_1 + \ldots + b_{\nu N}\chi_N \qquad (\nu = 0,1,\ldots,N)$$

are distinct from zero and satisfy the inequalities

$$\operatorname{ord} \Psi_0 < \operatorname{ord} \Psi_1 < \ldots < \operatorname{ord} \Psi_N .$$

There further exist $N+1$ constants $c_0, c_1, \ldots, c_N$ not all zero satisfying

$$c_0\Psi_0 + c_1\Psi_1 + \ldots + c_N\Psi_N = 0$$

because the $\Psi_\nu$ are $N+1$ linear forms in only $N$ series $\chi_1, \ldots, \chi_N$. Assume that, say

$$c_0 = \ldots = c_{\nu-1} = 0 \text{ , but } c_\nu \neq 0 .$$

Then evidently

$$\mathrm{ord}(c_0\Psi_0 + c_1\Psi_1 + \ldots + c_N\Psi_N) = \mathrm{ord}\ \Psi_\nu ,$$

and hence

$$c_0\Psi_0 + c_1\Psi_1 + \ldots + c_N\Psi_N \neq 0 ,$$

contrary to the hypothesis.

**45.** We wish to study *systems of homogeneous linear differential equations*

$$Q : w_h' = \sum_{k=1}^{m} q_{hk} w_k \qquad (h = 1,2,\ldots,m)$$

where, as already stated, the coefficients $q_{hk}$ are assumed to be *rational functions*, while the unknowns $w_1, \ldots, w_m$ may be any *series in* $K\langle z-c\rangle$. It is convenient to use matrix notation and to write

$$\underline{q} = \begin{bmatrix} q_{11} & \cdots & q_{1m} \\ \vdots & & \vdots \\ q_{m1} & \cdots & q_{mm} \end{bmatrix}, \quad \mathrm{w} = \begin{bmatrix} w_1 \\ \vdots \\ w_m \end{bmatrix}, \quad \mathrm{w}' = \begin{bmatrix} w_1' \\ \vdots \\ w_m' \end{bmatrix},$$

so that $Q$ can be written in the equivalent matrix form

$$Q : \mathrm{w}' = \underline{q}\mathrm{w} .$$

Every vector $\mathrm{w}$ satisfying this equation is called a *solution of* $Q$, or a *solution*, for short.

We are interested in the set $V_Q$ of *all* solutions $\mathrm{w}$ of $Q$. If $\mathrm{w}_1$ and $\mathrm{w}_2$ are solutions, and if $a_1$ and $a_2$ are constants, then the sum $a_1\mathrm{w}_1 + a_2\mathrm{w}_2$ evidently also is a solution. Thus $V_Q$ *is a vector space over* $K$.

Denote by $M$ the largest number of solutions of $Q$ that are linearly independent over $K\langle z-c\rangle$, so that

$$0 \leq M \leq m ,$$

because $m+1$ $m$-vectors with components in $K\langle z-c\rangle$ are always linearly dependent over this field.

If $M=0$, $V_Q$ consists only of the zero vector $\mathbf{0}$. Let us exclude this trivial case, and let us then denote by $\mathbf{w}_1, \ldots, \mathbf{w}_M$ any $M$ solutions of $Q$ that are linearly independent over $K\langle z-c\rangle$. Hence, if $\mathbf{w}$ is an arbitrary solution of $Q$, there exist $M+1$ elements $b, b_1, \ldots, b_M$ of $K\langle z-c\rangle$ that are not all zero and satisfy

$$b\mathbf{w} - (b_1\mathbf{w}_1+\ldots+b_N\mathbf{w}_N) = 0 ,$$

where necessarily $b \neq 0$. On dividing by $b$,

$$\mathbf{w} = a_1\mathbf{w}_1 + \ldots + a_M\mathbf{w}_M ,$$

and here $a_1, \ldots, a_M$ again are series. On differentiating this equation,

$$\mathbf{w}' = (a_1'\mathbf{w}_1+\ldots+a_M'\mathbf{w}_M) + (a_1\mathbf{w}_1'+\ldots+a_M\mathbf{w}_M') .$$

Now, by hypothesis,

$$\mathbf{w}' = \underline{\underline{g}}\mathbf{w} , \quad \mathbf{w}_1' = \underline{\underline{g}}\mathbf{w}_1, \ldots, \mathbf{w}_M' = \underline{\underline{g}}\mathbf{w}_M ,$$

whence

$$a_1'\mathbf{w}_1 + \ldots + a_M'\mathbf{w}_M = \mathbf{w}' - (a_1\mathbf{w}_1'+\ldots+a_M\mathbf{w}_M') = \underline{\underline{g}}[\mathbf{w}-(a_1\mathbf{w}_1+\ldots+a_M\mathbf{w}_M)] = 0 .$$

Since the vectors $\mathbf{w}_1, \ldots, \mathbf{w}_M$ are linearly independent, this implies that

$$a_1' = \ldots = a_M' = 0 ,$$

and so $a_1, \ldots, a_M$ are constants.

The result just proved may be expressed as follows.

(2): *Let $M$ be the maximal number of elements of $V_Q$ that are linearly independent over $K\langle z-c\rangle$, and let $\mathbf{w}_1, \ldots, \mathbf{w}_M$ be such independent elements. Then $\mathbf{w}_1, \ldots, \mathbf{w}_M$ form a basis of $V_Q$ over $K$, and the dimension of $V_Q$ over $K$ is equal to $M$.*

Thus, in particular, *the dimension of $V_Q$ over $K$ cannot exceed $m$.*

**46.** In addition to $V_Q$ and its subspaces we also require certain vector-spaces over $K(z)$. The elements of these new spaces are linear forms (functionals)

$$\lambda(\mathbf{w}) = \lambda = p_1w_1 + \ldots + p_mw_m$$

where the coefficients $p_1, \ldots, p_m$ are rational functions, while $w_1, \ldots, w_m$ denote the components of an arbitrary solution $\mathbf{w}$ of $Q$. Let $\Lambda$ be a set of such linear forms which constitutes a vector space over $K(z)$. Thus if $\lambda_1$ and $\lambda_2$ are in $\Lambda$, and if $a_1$ and $a_2$ are any two rational functions, then $a_1\lambda_1 + a_2\lambda_2$ also belongs to $\Lambda$. The dimension of $\Lambda$ over $K(z)$, $n$ say, obviously satisfies the inequality

$$0 \le n \le m .$$

In the special case when $n = 0$, $\Lambda$ consists only of the zero form

$$0 = 0w_1 + \ldots + 0w_m .$$

With every vector space $\Lambda$ so defined we associate that subset of $V_Q$, $V_Q(\Lambda)$ say, which consists of all those solutions $\mathbf{w}$ of $Q$ which have the property

$$\lambda(\mathbf{w}) = 0 \quad \text{for all} \quad \lambda \in \Lambda .$$

If $\mathbf{w}_1$ and $\mathbf{w}_2$ are in $V_Q(\Lambda)$, and if $b_1$ and $b_2$ are any constants, then $b_1\mathbf{w}_1 + b_2\mathbf{w}_2$ obviously also is an element of $V_Q(\Lambda)$. This set is thus a vector space over $K$ and is a vector subspace of $V_Q$.

We are mainly interested in a special class of vector spaces $\Lambda$. These are characterised by a certain closure property to which one comes in the following manner. Denote by $\kappa$ the least common denominator of the $m^2$ rational functions

$$q_{hk} \qquad (h,k = 1,2,\ldots,m)$$

and by $D$ the differential operator

$$D = \kappa \frac{d}{dz} .$$

If $\lambda$ and $\mathbf{w}$ are elements of $\Lambda$ and $V_Q$, respectively, evidently

$$D\lambda(\mathbf{w}) = \sum_{h=1}^{m} \kappa(p_h'w_h + p_h w_h') = \sum_{h=1}^{m} \kappa p_h' w_h + \sum_{h=1}^{m} \sum_{j=1}^{m} \kappa p_j q_{jh} w_h , \quad = \sum_{h=1}^{m} p_h^* w_h \quad \text{say,}$$

where we have put

$$p_h^* = \kappa\left\{p_h' + \sum_{j=1}^{m} p_j q_{jh}\right\} \qquad (h = 1,2,\ldots,m) .$$

The new linear form $D\lambda$ defined by

$$D\lambda = \sum_{h=1}^{m} p_h^* w_h$$

is called the *derivative of* $\lambda$, and the vector space $\Lambda$ is said to be *closed under derivation* if

$$\lambda \in \Lambda \text{ implies that also } D\lambda \in \Lambda .$$

The reason for including the factor $\kappa$ in the definition of $D$ will become clear later on.

**47.** Assume that $\Lambda$ is closed under derivation, and denote by $\lambda_1, \ldots, \lambda_n$ a basis of $\Lambda$. Thus every $\lambda$ in $\Lambda$ can be written in the form

$$\lambda = u_1\lambda_1 + \ldots + u_n\lambda_n$$

where $u_1, \ldots, u_n$ are rational functions. In order to exclude a trivial case, assume further that

$$M > n .$$

Since $\Lambda$ is closed under derivation, the new linear forms

$$D\lambda_1, \ldots, D\lambda_n$$

are again elements of $\Lambda$, hence can be written as

$$D\lambda_h = \sum_{k=1}^{n} \kappa u_{hk}\lambda_k \qquad (h = 1,2,\ldots,n)$$

where the coefficients $u_{hk}$ are certain rational functions.

Now, if $\mathbf{W}$ is an arbitrary solution of $Q$, put

$$\omega_h = \lambda_h(\mathbf{W}) \qquad (h = 1,2,\ldots,n) .$$

The $n$ series $\omega_1, \ldots, \omega_n$ so defined then satisfy the system of homogeneous linear differential equations

$$Q^* : \omega_h' = \frac{1}{\kappa} D\lambda_h(\mathbf{W}) = \sum_{k=1}^{n} u_{hk}\lambda_k(\mathbf{W}) = \sum_{k=1}^{n} u_{hk}\omega_k \qquad (h = 1,2,\ldots,n) .$$

On putting

$$\underline{\underline{u}} = \begin{bmatrix} u_{11} & \cdots & u_{1n} \\ \vdots & & \vdots \\ u_{n1} & \cdots & u_{nn} \end{bmatrix}, \quad \boldsymbol{\omega} = \begin{bmatrix} \omega_1 \\ \vdots \\ \omega_n \end{bmatrix}, \quad \boldsymbol{\omega}' = \begin{bmatrix} \omega_1' \\ \vdots \\ \omega_n' \end{bmatrix},$$

this system $Q^*$ can be written in matrix form as

$$Q^* : \boldsymbol{\omega}' = \underline{\underline{u}}\boldsymbol{\omega} .$$

In analogy to $V_Q$, denote by $V_{Q^*}$ the set of all solutions $\boldsymbol{\omega}$ of $Q^*$. Then

$V_{Q*}$ likewise is a vector space over $K$, say of dimension $N$. From (2) we deduce the basic inequality

$$0 \leq N \leq n < M .$$

Let $v_1, \ldots, v_M$ be an arbitrary basis of $V_Q$ over $K$, and put

$$\omega_{hk} = \lambda_h(v_k) \qquad \begin{pmatrix} h = 1,2,\ldots,n \\ k = 1,2,\ldots,M \end{pmatrix}$$

and

$$\boldsymbol{\omega}_k = \begin{bmatrix} \omega_{1k} \\ \vdots \\ \omega_{nk} \end{bmatrix} \qquad (k = 1,2,\ldots,M) .$$

These $M$ vectors are then solutions of $Q^*$, and so at most $n$ of them can be linearly independent over $K$. It follows that there exists a constant $M \times (M-n)$ matrix

$$\underline{\underline{a}}^* = \begin{bmatrix} a_{11} & \cdots & a_{1,M-n} \\ \vdots & & \vdots \\ a_{M1} & \cdots & a_{M,M-n} \end{bmatrix}$$

of exact rank $M - n$ such that $\boldsymbol{\omega}_1, \ldots, \boldsymbol{\omega}_M$ are connected by the $M - n$ independent linear equations

$$\boldsymbol{\omega}_1 a_{1k} + \ldots + \boldsymbol{\omega}_M a_{Mk} = 0 \qquad (k = 1,2,\ldots,M-n) .$$

The matrix $\underline{\underline{a}}^*$ can be completed (in many ways) to a constant non-singular square matrix

$$\underline{\underline{a}} = \begin{bmatrix} a_{11} & \cdots & a_{1M} \\ \vdots & & \vdots \\ a_{M1} & \cdots & a_{MM} \end{bmatrix}, \text{ where } \det \underline{\underline{a}} \neq 0 .$$

Now put

$$w_k = v_1 a_{1k} + \ldots + v_M a_{Mk} \qquad (k = 1,2,\ldots,M) ;$$

since $\underline{\underline{a}}$ is non-singular, $w_1, \ldots, w_M$ form again a basis of $V_Q$. Furthermore,

$$\lambda_h(w_k) = \omega_{h1} a_{1k} + \ldots + \omega_{hM} a_{Mk}$$

is the $h$th component of the vector

$$\boldsymbol{\omega}_1 a_{1k} + \ldots + \boldsymbol{\omega}_M a_{Mk}$$

and so is equal to zero for $h = 1, 2, \ldots, n$ and $k = 1, 2, \ldots, M-n$. This establishes the following result.

**(3):** *Let the dimensions* $M$ *of* $V_Q$ *over* $K$, *and* $n$ *of* $\Lambda$ *over* $K(z)$, *satisfy the inequality*

$$M > n,$$

*and further let* $\Lambda$ *be closed under derivation.*

*Then there exists a basis* $\mathbf{w}_1, \ldots, \mathbf{w}_M$ *of* $V_Q$ *over* $K$ *with the property that*

$$\lambda(\mathbf{w}_k) = 0 \quad \textit{for} \quad k = 1, 2, \ldots, M-n \quad \textit{and for all} \quad \lambda \in \Lambda .$$

**48.** In the immediately following sections, the system $Q$ will be restricted by assuming that

$$\text{ord}\, q_{hk} \geq 0 \qquad\qquad (h,k = 1,2,\ldots,m),$$

so that, in particular, the least common denominator $\kappa$ has the order

$$\text{ord}\, \kappa = 0 .$$

This is equivalent to demanding that *none of the rational functions* $q_{hk}$ *has a pole at* $z = c$. We say that the system $Q$ is *regular* if this property holds.

Since $Q$ is regular, the matrix $\underline{q}$ can be written as a formal power series

$$\underline{q} = \sum_{l=0}^{\infty} \underline{q}_{(l)}(z-c)^{l},$$

where the coefficients $\underline{q}_{(l)}$ are constant matrices. If $\mathbf{w}$ is an arbitrary solution of $Q$, put

$$\zeta = \min(\text{ord}\, w_1,\ldots,\text{ord}\, w_m), \quad = \text{ord}\, w_1 \quad \text{say.}$$

Here

$$\zeta \geq 0 .$$

For if $\zeta$ were negative, we should have

$$\text{ord}\, w_1' = \zeta - 1, \text{ but } \text{ord}\left(\sum_{k=1}^{m} q_{1k}w_k\right) \geq \zeta,$$

giving a contradiction.

We can also write $\mathbf{w}$ and $\mathbf{w}'$ as power series

$$\mathbf{w} = \sum_{l=0}^{\infty} \mathbf{w}_{(l)}(z-c)^{l} \quad \text{and} \quad \mathbf{w}' = \sum_{l=0}^{\infty} (l+1)\mathbf{w}_{(l+1)}(z-c)^{l},$$

where $w_{(0)}, w_{(1)}, w_{(2)}, \ldots$ are certain constant vectors. The first of these vectors, $w_{(0)}$, is called the *constant part* of $w$.

On substituting these series for $\underline{q}$, $w$, and $w'$ in the matrix equation

$$Q : w' = \underline{q}w$$

and comparing the coefficients of equal powers of $z - c$ on both sides, we find that

$$(l+1)w_{(l+1)} = \underline{q}_{(0)}w_{(l)} + \underline{q}_{(1)}w_{(l-1)} + \ldots + \underline{q}_{(l)}w_{(0)} \qquad (l = 0,1,2,3,\ldots) .$$

These recursive formulae allow us to express successively $w_{(1)}, w_{(2)}, w_{(3)}, \ldots$ in terms of the matrices $\underline{q}_{(l)}$ and the vector $w_{(0)}$. The result has the form

$$w_{(l)} = \underline{q}_{[l]}w_{(0)} \qquad (l = 1,2,3,\ldots)$$

where, for each suffix $l$, $\underline{q}_{[l]}$ is a certain constant matrix that depends only on $\underline{q}$. In the lowest cases,

$$\underline{q}_{[1]} = \underline{q}_{(0)} , \quad \underline{q}_{[2]} = \tfrac{1}{2}\underline{q}_{(0)}^2 + \tfrac{1}{2}\underline{q}_{(1)} , \quad \underline{q}_{[3]} = \tfrac{1}{6}\underline{q}_{(0)}^3 + \tfrac{1}{6}\underline{q}_{(0)}\underline{q}_{(1)} + \tfrac{1}{3}\underline{q}_{(1)}\underline{q}_{(0)} + \tfrac{1}{3}\underline{q}_{(2)} ,$$

and so forth.

Let $\underline{E}$ be the $m \times m$ unit matrix, and put

$$\underline{v} = \underline{E} + \sum_{l=1}^{\infty} \underline{q}_{[l]}(z-c)^l .$$

There is then exactly one solution

$$w = w_{(0)} + w_{(1)}(z-c) + w_{(2)}(z-c)^2 + \ldots$$

of $Q$ which has the given constant part $w_{(0)}$, and this solution has the explicit form

$$w = \underline{v}w_{(0)} .$$

**49.** In the formula just proved choose for $w_{(0)}$ successively each of the columns of the unit matrix $\underline{E}$. Then $w$ becomes successively identical with the columns of the matrix $\underline{v}$, the columns $v_1, \ldots, v_m$ say.

From the definition of $\underline{v}$,

$$\det \underline{v} = 1 + \sum_{l=1}^{\infty} \upsilon_l(z-c)^l$$

where $\upsilon_1, \upsilon_2, \upsilon_3, \ldots$ are certain constants. Therefore $\underline{v}$ is non-singular and in addition satisfies the equation

$$\mathrm{ord}(\det \underline{\underline{v}}) = 0 \ .$$

It follows that the $m$ solutions $v_1, \ldots, v_m$ of $Q$ are linearly independent over the field $K\langle z-c\rangle$ and therefore form a basis of $V_Q$ over $K$. Thus

(4): *When $Q$ is regular, $V_Q$ has dimension $m$ over $K$.*

The most general basis $w_1, \ldots, w_m$ of $V_Q$ is now given by

$$w_k = \sum_{h=1}^{m} v_h a_{hk} \qquad (k = 1,2,\ldots,m)$$

where

$$\underline{\underline{a}} = \begin{bmatrix} a_{11} & \cdots & a_{1m} \\ \vdots & & \vdots \\ a_{m1} & \cdots & a_{mm} \end{bmatrix}$$

is any non-singular constant matrix,

$$\det \underline{\underline{a}} \neq 0$$

The product matrix

$$\underline{\underline{w}} = \underline{\underline{va}} \ ,$$

of which $w_1, \ldots, w_m$ are the columns, likewise is non-singular, and in fact

$$\mathrm{ord}(\det \underline{\underline{w}}) = 0 \ .$$

**50.** Let $Q$ again be regular, and let

$$f = \begin{bmatrix} f_1 \\ \vdots \\ f_m \end{bmatrix} \neq 0$$

be a fixed solution of $Q$. The *rank of* f is defined as the largest integer $\rho$ such that some $\rho$ components of f are linearly independent over $K(z)$. From this definition, $1 \leq \rho \leq m$.

We assume for the moment that the stronger inequality

$$1 \leq \rho \leq m-1$$

holds, that is, that *the components of* f *are linearly dependent over* $K(z)$. If, say, the components $f_{h_1}, \ldots, f_{h_\rho}$ are linearly independent over $K(z)$, then the remaining $m - \rho$ components of f can be written as linear forms in $f_{h_1}, \ldots, f_{h_\rho}$

with coefficients which are rational functions.

It follows that there exist $m - \rho$ , *but no more*, linear forms

$$\lambda_h(\mathrm{w}) = \lambda_h = p_{h1}w_1 + \dots + p_{hm}w_m \qquad (h = 1,2,\dots,m-\rho)$$

which are linearly independent over, and have coefficients in, the field $K(z)$ , such that

$$\lambda_h(\mathrm{f}) = 0 \text{ and hence also } D\lambda_h(\mathrm{f}) = 0 \qquad (h = 1,2,\dots,m-\rho) .$$

The set $\Lambda$ of all linear forms $\lambda$ satisfying

$$\lambda(\mathrm{f}) = 0$$

is then a vector space over $K(z)$ of dimension $m - \rho$ , and it is closed under derivation. By the earlier lemma (3), this implies the following result.

(5): *If $Q$ is regular, and if $\mathrm{f} \neq 0$ is a solution of $Q$ of rank $\rho < m$ , then there exists a basis $\mathrm{w}_1, \dots, \mathrm{w}_m$ of $V_Q$ such that*

$\lambda(\mathrm{w}_k) = 0$ *for* $k = 1, 2, \dots, \rho$ *and for all* $\lambda$ *satisfying* $\lambda(\mathrm{f}) = 0$ .

51. Let $\mathrm{w}_1, \dots, \mathrm{w}_m$ be the basis in (5), and let further

$$\underline{\mathrm{w}} = (\mathrm{w}_1,\dots,\mathrm{w}_m) = \begin{bmatrix} w_{11} & \cdots & w_{1m} \\ \vdots & & \vdots \\ w_{m1} & \cdots & w_{mm} \end{bmatrix}$$

be the square matrix of which these vectors form the columns. Since $Q$ is regular, by the proof in §48,

$$\operatorname{ord} w_{hk} \geq 0 \qquad (h,k = 1,2,\dots,m)$$

and

$$\operatorname{ord}(\det \underline{\mathrm{w}}) = 0 .$$

Denote now by $\pi = (i_1,\dots,i_\rho,l_1,\dots,l_{m-\rho})$ any permutation of the set $1, 2, \dots, m$ for which

$$i_1 < i_2 < \dots < i_\rho , \quad l_1 < l_2 < \dots < l_{m-\rho} .$$

With this permutation we associate the pair of complementary minors

$$\Delta(\pi) = \begin{vmatrix} w_{i_1 1} & \cdots & w_{i_1 \rho} \\ \vdots & & \vdots \\ w_{i_\rho 1} & \cdots & w_{i_\rho \rho} \end{vmatrix} \quad \text{and} \quad \Delta^*(\pi) = \begin{vmatrix} w_{l_1,\rho+1} & \cdots & w_{l_1,m} \\ \vdots & & \vdots \\ w_{l_{m-\rho},\rho+1} & \cdots & w_{l_{m-\rho},m} \end{vmatrix}$$

of $\underline{w}$ . Then evidently

$$\operatorname{ord} \Delta(\pi) \geq 0 \quad \text{and} \quad \operatorname{ord} \Delta^*(\pi) \geq 0 \quad \text{for all } \pi ,$$

and, by the Laplace development of the determinant,

$$\det \underline{w} = \sum \mp \Delta(\pi)\Delta^*(\pi)$$

where the summation extends over all permutations $\pi$ . It follows that there is at least one permutation $\pi$ for which

$$\operatorname{ord}\left(\Delta(\pi)\Delta^*(\pi)\right) = 0$$

and hence also

$$\Delta(\pi) \neq 0 \quad \textit{and even} \quad \operatorname{ord} \Delta(\pi) = 0 .$$

This property implies, firstly, that the components $f_{i_1}, \ldots, f_{i_\rho}$ of f are linearly independent over $K(z)$ . For otherwise there would be $\rho$ rational functions $p_{i_1}, \ldots, p_{i_\rho}$ not all zero satisfying

$$p_{i_1} f_{i_1} + \ldots + p_{i_\rho} f_{i_\rho} = 0 ,$$

and then, by (5), we should also have

$$p_{i_1} w_{i_1,k} + \ldots + p_{i_\rho} w_{i_\rho,k} = 0 \qquad (k = 1,2,\ldots,\rho) ,$$

and so it would follow that $\Delta(\pi) = 0$ .

Secondly, because $f_{i_1}, \ldots, f_{i_\rho}$ are linearly independent over $K(z)$ , and since f has the rank $\rho$ , there exists a set of $\rho(m-\rho)$ rational functions $d_{hj}$ such that

$$f_{l_j} = \sum_{h=1}^{\rho} d_{hj} f_{i_h} \qquad (j = 1,2,\ldots,m-\rho) .$$

Once more by (5), we have then also

$$w_{l_j,k} = \sum_{h=1}^{\rho} d_{hj} w_{i_h,k} \qquad \begin{pmatrix} j = 1,2,\ldots,m-\rho \\ k = 1,2,\ldots,\rho \end{pmatrix} .$$

For each fixed value of $j$ , and for $k = 1, 2, \ldots, \rho$ , this is a system of $\rho$ linear equations for the $\rho$ rational functions

$$d_{1j}, \ldots, d_{\rho j} ,$$

with the determinant $\Delta(\pi) \neq 0$ . These rational functions have therefore the explicit values

$$d_{hj} = \frac{\Delta_{hj}}{\Delta(\pi)} \qquad \begin{pmatrix} h = 1,2,\dots,\rho \\ j = 1,2,\dots,m-\rho \end{pmatrix} .$$

Here $\Delta_{hj}$ denotes that determinant which is obtained from $\Delta(\pi)$ if the row

$$w_{i_h,1}, \dots, w_{i_h,\rho}$$

is replaced by the new row

$$w_{l_j,1}, \dots, w_{l_j,\rho} .$$

Thus also the elements of $\Delta_{hj}$ have non-negative orders, and this implies that

$$\text{ord } \Delta_{hj} \geq 0 .$$

Since $\text{ord } \Delta(\pi) = 0$ , it follows then finally that

$$\text{ord } d_{hj} \geq 0 \qquad \begin{pmatrix} h = 1,2,\dots,\rho \\ j = 1,2,\dots,m-\rho \end{pmatrix} .$$

52. Next denote by $W_Q$ the subset of $V_Q$ which consists of all those solutions $\mathbf{w}$ of $Q$ , the components of which satisfy the relations

(6): $$w_{l_j} = \sum_{h=1}^{\rho} d_{hj} w_{i_h} \qquad (j = 1,2,\dots,m-\rho) .$$

It is obvious that $W_Q$ likewise is a vector space over $K$ . By §51, the vectors $\mathbf{f}$ and $\mathbf{w}_1, \dots, \mathbf{w}_\rho$ belong to $W_Q$ .

With each vector $\mathbf{w}$ in $W_Q$ we associate now the $\rho$-vector

$$\mathbf{w}^0 = \begin{bmatrix} w_{i_1} \\ \vdots \\ w_{i_\rho} \end{bmatrix} .$$

In particular, we denote by $\mathbf{f}^0, \mathbf{w}_1^0, \dots, \mathbf{w}_\rho^0$ the $\rho$-vectors that are by this definition associated with the vectors $\mathbf{f}, \mathbf{w}_1, \dots, \mathbf{w}_\rho$ .

By hypothesis, the components of $\mathbf{w}^0$ satisfy the differential equations

$$w'_{i_h} = \sum_{k=1}^{m} q_{i_h,k} w_k \qquad (h = 1,2,\dots,\rho)$$

which form a part of the system $Q$ . By means of the relations (6) we can here eliminate the components $w_{l_1}, \ldots, w_{l_{m-\rho}}$, and then find that

$$w'_{i_h} = \sum_{k=1}^{\rho} q_{i_h i_k} w_{i_k} + \sum_{j=1}^{m-\rho} \sum_{k=1}^{\rho} q_{i_h l_j} d_{kj} w_{i_k} \,, = \sum_{k=1}^{\rho} q^0_{hk} w_{i_k} \quad \text{say} \qquad (h = 1,2,\ldots,\rho) \,,$$

where we have put

$$q^0_{hk} = q_{i_h i_k} + \sum_{j=1}^{m-\rho} q_{i_h l_j} d_{kj} \qquad (h = 1,2,\ldots,\rho) \,.$$

This proves that the $\rho$-vector $\mathbf{w}^0$ satisfies a system of $\rho$ homogeneous linear differential equations

$$Q^0 : w'_{i_h} = \sum_{k=1}^{\rho} q^0_{hk} w_{i_k} \qquad (h = 1,2,\ldots,\rho) \,,$$

where the new coefficients $q^0_{hk}$ are again rational functions. Moreover, since

$$\text{ord } q_{hk} \geq 0 \quad \text{and} \quad \text{ord } d_{hj} \geq 0$$

for all suffixes $h$, $j$, and $k$, these coefficients satisfy the inequalities

$$\text{ord } q^0_{hk} \geq 0 \qquad (h,k = 1,2,\ldots,\rho) \,,$$

and hence the new system $Q^0$ is likewise regular. The result obtained may be formulated as follows.

THEOREM 5. *Let*

$$Q : w'_h = \sum_{k=1}^{m} q_{hk} w_k \qquad (h = 1,2,\ldots,m)$$

*be a regular system of homogeneous linear differential equations, and let* $\mathbf{f} \neq 0$ *be a solution of* $Q$ *of rank* $\rho < m$. *Then it is possible to select* $\rho$ *components* $f_{i_1}, \ldots, f_{i_\rho}$ *of* $\mathbf{f}$ *which have the following properties.*

*(a):* $f_{i_1}, \ldots, f_{i_\rho}$ *are linearly independent over* $K(z)$.

*(b): The* $\rho$*-vector* $\mathbf{f}^0$ *formed by these components is a solution of a second regular system*

$$Q^0 : w'_{i_h} = \sum_{k=1}^{\rho} q^0_{hk} w_{i_k} \qquad (h = 1,2,\ldots,\rho)$$

*of homogeneous linear differential equations, likewise with coefficients* $q^0_{hk}$ *in* $K(z)$.

This theorem is due to Shidlovski. It proves that it suffices to study those solutions of a regular system $Q$ which have *maximal* rank $\rho = m$ .

**53.** The remainder of this chapter deals with a certain determinant associated with a general linear form $\lambda$ and its derived forms.

As before, let

$$Q : w'_h = \sum_{k=1}^{m} q_{hk} w_k \qquad (h = 1,2,\ldots,m)$$

be a fixed system of homogeneous linear differential equations with coefficients $q_{hk}$ in $K(z)$ . As in §46, we denote by $\kappa$ the least common denominator of these $m^2$ coefficients and by $D$ the differential operator

$$D = \kappa \frac{d}{dz} .$$

From now on the constant $c$ defining the field $K\langle z-c\rangle$ may be *variable*. Therefore, for the sake of clarity, we shall write

$$\mathrm{ord}_c\, f$$

for the order of an element of $K\langle z-c\rangle$ . For the present, $Q$ is assumed to be *regular relative to* $K\langle z-c\rangle$ ; this means as before that

$$\mathrm{ord}_c\, q_{hk} \geq 0 \qquad (h,k = 1,2,\ldots,m) ,$$

or equivalently that

$$\mathrm{ord}_c\, \kappa = 0 .$$

A solution $\mathbf{w}$ of $Q$ naturally now has components in the variable field $K\langle z-c\rangle$ .

Denote by

$$\lambda(\mathbf{w}) = \lambda = p_1 w_1 + \ldots + p_m w_m$$

an arbitrary linear form with coefficients $p_1, \ldots, p_m$ in $K(z)$ that do not all vanish. We derive from $\lambda$ an important infinite sequence of new forms

$$\lambda_h(\mathbf{w}) = \lambda_h = p_{h1} w_1 + \ldots + p_{hm} w_m \qquad (h = 1,2,3,\ldots)$$

with coefficients $p_{hk}$ in $K(z)$ by applying again and again the operator $D$ ; thus

$$\lambda_1(\mathbf{w}) = \lambda(\mathbf{w}) , \quad \lambda_{h+1}(\mathbf{w}) = D\lambda_h(\mathbf{w}) \qquad (h = 1,2,3,\ldots) .$$

For the coefficients of these linear forms this is equivalent to the relations

(7): $p_{1k} = p_k$ for all $k$ , and $p_{h+1,k} = \kappa\left(p'_{hk} + \sum_{j=1}^{m} p_{hj} q_{jk}\right)$ for all $h$ and $k$ .

These formulae make it plain that *the* $p_{hk}$ *do not in any way depend on the choice of the constant* $c$ .

Since the $\lambda_h$ are linear forms in only $m$ letters $w_1, \ldots, w_m$ , not more than $m$ of them can be linearly independent over $K(z)$ . There exists then also an integer $\mu$ satisfying

$$1 \le \mu \le m$$

such that $\lambda_1, \ldots, \lambda_\mu$ are linearly independent over $K(z)$ , while the immediately following form $\lambda_{\mu+1}$ can be written as

(8): $$\lambda_{\mu+1} = b_1\lambda_1 + \ldots + b_\mu\lambda_\mu ,$$

where $b_1, \ldots, b_\mu$ are certain rational functions. The number $\mu$ so defined is called the *rank of* $\lambda$ . For the present *this rank* $\mu$ *is assumed to be strictly less than* $m$ .

Denote by $\Lambda$ the linear vector space over $K(z)$ with the basis $\lambda_1, \ldots, \lambda_\mu$ , so that $\Lambda$ consists of all linear forms

$$a_1\lambda_1 + \ldots + a_\mu\lambda_\mu$$

where $a_1, \ldots, a_\mu$ are rational functions. From the definition of the forms $\lambda_h$ and from the equation (8) it is obvious that $\Lambda$ *is closed under derivation and has the dimension* $\mu$ *over* $K(z)$ .

**54.** Lemma (3) can now be applied again, with $M = m$ and $n = \mu$ . By this lemma, there exists a basis $\mathrm{w}_1, \ldots, \mathrm{w}_m$ of $V_Q$ over $K$ (which is not necessarily the same as that in §§50-52) such that

(9): $$\lambda_h(\mathrm{w}_k) = 0 \quad \text{for } h = 1, 2, \ldots, \mu \text{ and } k = 1, 2, \ldots, m-\mu .$$

The following construction allows us to replace these equations by simpler ones which are equivalent.

The $\mu \times m$ matrix

$$\underline{p}^* = \begin{bmatrix} p_{11} & \cdots & p_{1m} \\ \vdots & & \vdots \\ p_{\mu 1} & \cdots & p_{\mu m} \end{bmatrix}$$

has the exact rank $\mu$ because $\lambda_1, \ldots, \lambda_\mu$ are linearly independent. Therefore, after possibly applying a trivial change of notation, we may without loss of generality assume that the minor

$$P = \begin{vmatrix} p_{11} & \cdots & p_{1\mu} \\ \vdots & & \vdots \\ p_{\mu 1} & \cdots & p_{\mu\mu} \end{vmatrix}$$

of $\underline{p}^*$ does not vanish.

It follows that there exists a set of $\mu(m-\mu)$ rational functions

$$e_{ij}$$

such that

$$p_{hj} = \sum_{i=1}^{\mu} p_{hi} e_{ij} \qquad \begin{pmatrix} h = 1,2,\ldots,\mu \\ j = \mu+1,\mu+2,\ldots,m \end{pmatrix} .$$

From their definition, *these functions* $e_{ij}$ *are independent of the choice of the constant* $c$ . Further, for $h = 1, 2, \ldots, \mu$ and $k = 1, 2, \ldots, m-\mu$ ,

$$\lambda_h(w_k) = \sum_{i=1}^{\mu} p_{hi} w_{ik} + \sum_{j=\mu+1}^{m} p_{hj} w_{jk} = \sum_{i=1}^{\mu} p_{hi} \left( w_{ik} + \sum_{j=\mu+1}^{m} e_{ij} w_{jk} \right) = \sum_{i=1}^{\mu} p_{hi} W_{ik} ,$$

where we have put

$$W_{ik} = w_{ik} + \sum_{j=\mu+1}^{m} e_{ij} w_{jk} .$$

It follows then from (9) that

$$\lambda_h(w_k) = \sum_{i=1}^{\mu} p_{hi} W_{ik} = 0 \qquad \begin{pmatrix} h = 1,2,\ldots,\mu \\ k = 1,2,\ldots,m-\mu \end{pmatrix} .$$

For each value of $k$ , this is a system of $\mu$ homogeneous linear equations with the determinant $P \neq 0$ for the $\mu$ unknowns $W_{1k}, \ldots, W_{\mu k}$ . Hence

$$W_{ik} = 0 \quad \text{for} \quad i = 1, 2, \ldots, \mu \quad \text{and} \quad k = 1, 2, \ldots, m-\mu ,$$

which is equivalent to the formulae

$$(10): \qquad w_{ik} + \sum_{j=\mu+1}^{m} e_{ij} w_{jk} = 0 \quad \text{for} \quad i = 1, 2, \ldots, \mu \quad \text{and} \quad k = 1, 2, \ldots, m-\mu .$$

55. The $\mu(m-\mu)$ linear equations (10) split up into $\mu$ separate systems of $m - \mu$ equations for the $\mu$ sets of $m - \mu$ unknowns $e_{i,\mu+1}, e_{i,\mu+2}, \ldots, e_{im}$ . These systems of linear equations all have the same coefficient matrix

$$\underline{\underline{w}}^{(0)} = \begin{bmatrix} w_{\mu+1,1} & w_{\mu+1,2} & \cdots & w_{\mu+1,m-\mu} \\ w_{\mu+2,1} & w_{\mu+2,2} & \cdots & w_{\mu+2,m-\mu} \\ \vdots & \vdots & & \vdots \\ w_{m1} & w_{m2} & \cdots & w_{m,m-\mu} \end{bmatrix} .$$

We prove now that

(11):
$$\det \underline{\underline{w}}^{(0)} \neq 0 .$$

For this purpose consider again the matrix

$$\underline{\underline{w}} = \begin{bmatrix} w_{11} & \cdots & w_{1m} \\ \vdots & & \vdots \\ w_{m1} & \cdots & w_{mm} \end{bmatrix} .$$

For each pair of suffixes $i = 1, 2, \ldots, \mu$ and $j = \mu+1, \mu+2, \ldots, m$ multiply in $\underline{\underline{w}}$ the $j$th row by the factor $e_{ij}$ and add to the $i$th row. By (10), the effect of all these operations is to change $\underline{\underline{w}}$ into a new matrix of the form

$$\underline{\underline{w}}^{+} = \begin{bmatrix} \underline{\underline{0}} & \underline{\underline{w}}^{(1)} \\ \underline{\underline{w}}^{(0)} & \underline{\underline{w}}^{(2)} \end{bmatrix} ,$$

where $\underline{\underline{0}}$, $\underline{\underline{w}}^{(1)}$, and $\underline{\underline{w}}^{(2)}$ denote the $\mu \times (m-\mu)$ matrix consisting only of zeros, a certain $\mu \times \mu$ matrix, and a certain $(m-\mu) \times \mu$ matrix, respectively. It follows then by the Laplace development of $\det \underline{\underline{w}}^{+}$ that

$$\det \underline{\underline{w}} = \det \underline{\underline{w}}^{+} = \pm \det \underline{\underline{w}}^{(0)} . \det \underline{\underline{w}}^{(1)} ,$$

and this implies the inequality (11) because $\det \underline{\underline{w}} \neq 0$ .

56. Since the determinant

$$\Omega = \det \underline{\underline{w}}^{(0)}$$

does not vanish, the equations (10) can be solved for the $e_{ij}$ in the form

(12):
$$e_{ij} = -\frac{\Omega_{ij}}{\Omega} \qquad \begin{pmatrix} i = 1,2,\ldots,\mu \\ j = \mu+1,\mu+2,\ldots,m \end{pmatrix} ,$$

where $\Omega_{ij}$ is that determinant which is obtained from $\Omega$ if the row

$$w_{j1}, w_{j2}, \ldots, w_{j,m-\mu}$$

of $\Omega$ is replaced by the new row

$$w_{i1}, w_{i2}, \ldots, w_{i,m-\mu} .$$

From this representation of $e_{ij}$ we can now deduce the following important property of these rational functions. (See §43 for the definition of $\nabla$ .)

(13): *There exists a positive integer* $C_0$ *which depends only on the system* $Q$ *, but is independent of* $c$ *and of the linear form* $\lambda$ *, such that if* $\mu < m$ *, then*

$$\nabla(e_{ij}) \leq C_0 \quad \textit{for all } i \textit{ and } j .$$

Proof. For a fixed value of $c$ , let

$$\phi_1, \phi_2, \ldots, \phi_s$$

be the set of all minors of all orders $1, 2, \ldots, m$ of the matrix $\underline{v}$ constructed in §48, arranged in some fixed but arbitrary order. Since $\det \underline{v} \neq 0$ , these minors cannot all vanish simultaneously; all of them are series in $K\langle z-c\rangle$ .

The matrix $\underline{w}$ has the form

$$\underline{w} = \underline{va} ,$$

where $\underline{a}$ is a certain constant non-singular matrix. Therefore every minor of $\underline{w}$ can be written as a sum

$$b_1\phi_1 + \ldots + b_s\phi_s$$

with constant coefficients $b_1, \ldots, b_s$ . This is in particular true for the special minors $\Omega$ and $\Omega_{ij}$ of $\underline{w}$ which can thus be expressed in the form

$$\Omega = c_1\phi_1 + \ldots + c_s\phi_s \quad \text{and} \quad -\Omega_{ij} = c_{ij1}\phi_1 + \ldots + c_{ijs}\phi_s \qquad \begin{pmatrix} i = 1,2,\ldots,\mu \\ j = \mu+1,\mu+2,\ldots,m \end{pmatrix} ,$$

where also the coefficients $c_\sigma$ and $c_{ij\sigma}$ are constants.

Denote by $t$ , where $1 \leq t \leq s$ , the largest number of series $\phi_1, \ldots, \phi_s$ that are linearly independent over $K(z)$ and, without loss of generality, let $\phi_1, \ldots, \phi_t$ be such independent series. If $t < s$ , the remaining series $\phi_{t+1}, \ldots, \phi_s$ can now be written in the form

$$\phi_\sigma = \sum_{\tau=1}^{t} g_{\sigma\tau}\phi_\tau \qquad (\sigma = t+1,t+2,\ldots,s) ,$$

where the new coefficients $g_{\sigma\tau}$ are rational functions that depend on $Q$ and on $c$ , but do not depend on $\lambda$ . It follows that

$$\Omega = \sum_{\tau=1}^{t} c_\tau\phi_\tau + \sum_{\sigma=t+1}^{s} c_\sigma \sum_{\tau=1}^{t} g_{\sigma\tau}\phi_\tau = \sum_{\tau=1}^{t} \left(c_\tau + \sum_{\sigma=t+1}^{s} c_\sigma g_{\sigma\tau}\right)\phi_\tau$$

and similarly also that

$$-\Omega_{ij} = \sum_{\tau=1}^{t} \left(c_{ij\tau} + \sum_{\sigma=t+1}^{s} c_{ij\sigma} g_{\sigma\tau}\right)\phi_\tau \qquad \begin{pmatrix} i = 1,2,\ldots,\mu \\ j = \mu+1,\mu+2,\ldots,m \end{pmatrix},$$

whence

$$(14): \qquad e_{ij} = \left\{\sum_{\tau=1}^{t} \left(c_{ij\tau} + \sum_{\sigma=t+1}^{s} c_{ij\sigma} g_{\sigma\tau}\right)\phi_\tau\right\} \left\{\sum_{\tau=1}^{t} \left(c_\tau + \sum_{\sigma=t+1}^{s} c_\sigma g_{\sigma\tau}\right)\phi_\tau\right\}^{-1}.$$

Since $\Omega \neq 0$, at least one of the sums

$$c_\tau + \sum_{\sigma=t+1}^{s} c_\sigma g_{\sigma\tau} \qquad (\tau = 1,2,\ldots,t)$$

must be distinct from zero. Without loss of generality, let this be the sum which belongs to the suffix $\tau = 1$.

The series $\phi_1, \ldots, \phi_t$ are linearly independent over $K(z)$, and the $e_{ij}$ are elements of $K(z)$. The identities (14) therefore imply that the $e_{ij}$ allow the simplified representations

$$e_{ij} = \left(c_{ij1} + \sum_{\sigma=t+1}^{s} c_{ij\sigma} g_{\sigma 1}\right) \left(c_1 + \sum_{\sigma=t+1}^{s} c_\sigma g_{\sigma 1}\right)^{-1}.$$

Any alteration of the polynomials $p_1, \ldots, p_m$ in $\lambda$ affects the $e_{ij}$ only to the extent that, if $\mu$ remains fixed, the coefficients $c_{ij\sigma}$ and $c_\sigma$ in this formula are replaced by other constants, while the rational functions $g_{\sigma 1}$ remain unchanged. Thus the degrees of both the numerators and the denominators of the $e_{ij}$ remain bounded, and hence the degrees of the rational functions $e_{ij}$ themselves also cannot exceed a certain positive integer $C_1$ which depends on $Q$ and $c$, but not on $\lambda$. In fact, since the $e_{ij}$ are independent of $c$, $C_1$ may likewise be chosen independent of $c$.

This proof can be applied trivially also when $t = s$, and it then shows that the $e_{ij}$ are constants. Hence the assertion still remains valid.

**57.** From now on *we no longer demand that* $Q$ *be regular*, but we specialize the linear form $\lambda$ by assuming that *its coefficients* $p_1, \ldots, p_m$ *are polynomials*.

The coefficients $p_{hk}$ of the derived forms $\lambda_h$ are given by

$$(15): \; p_{1k} = p_k \text{ for all } k, \text{ and } p_{h+1,k} = \kappa p'_{hk} + \sum_{j=1}^{m} p_{hj}(\kappa q_{jk}) \text{ for all } h \text{ and } k.$$

It follows at once that also *the* $p_{hk}$ *are polynomials*, because so are $\kappa$ and the $\kappa q_{hk}$ .

Upper estimates for the degrees of the $p_{hk}$ are obtained as follows. Put

$$C_2 = \max_{h,k} \{\partial(\kappa), \partial(\kappa q_{hk})\}$$

and

$$X = \max\{\partial(p_1),\ldots,\partial(p_m)\} \ , \quad X_h = \max\{\partial(p_{h1}),\ldots,\partial(p_{hm})\} \qquad (h = 1,2,3,\ldots) \ .$$

Then, from (15),

$$X_1 = X \ , \text{ and } X_{h+1} \le X_h + C_2 \qquad (h = 1,2,3,\ldots)$$

and hence

$$X_h \le X + (h-1)C_2 \qquad (h = 1,2,3,\ldots) \ .$$

Again let $\lambda$ be of rank $\mu$ and choose the notation so that the minor

$$P = \begin{vmatrix} p_{11} & \cdots & p_{1\mu} \\ \vdots & & \vdots \\ p_{\mu 1} & \cdots & p_{\mu\mu} \end{vmatrix}$$

does not vanish,

$$P \neq 0 \ .$$

This minor is a polynomial of degree

(16): $$\partial(P) \le \sum_{h=1}^{\mu} \{X+(h-1)C_2\} = \mu X + \frac{\mu(\mu-1)}{2} C_2 \ ,$$

and being distinct from zero, it satisfies for every constant $c$ the order relation

(17): $$\mathrm{ord}_c \, P \le \mu X + \frac{\mu(\mu-1)}{2} C_2 \ .$$

**58.** Since $Q$ need no longer be regular in $K(z-c)$ , the vector space $V_Q$ relative to this field will now in general no longer have the dimension $m$ over $K$ and might actually consist only of the zero vector. However, let us assume that its dimension is at least $1$ , and then let $\mathbf{f} \neq 0$ be a fixed solution of $Q$ with components in $K(z-c)$ and of *maximal* possible rank $\rho = m$ .

Let $\lambda$ be the same linear form as in §57, so that in particular $\mu \le m - 1$ . The values

$$\lambda_h(\mathbf{f}) = p_{h1}f_1 + \ldots + p_{hm}f_m \qquad (h = 1,2,\ldots,\mu)$$

of the derived linear forms $\lambda_h$ at $\mathbf{w} = \mathbf{f}$ can be given a more convenient form by means of the previous identities

$$p_{hj} = \sum_{i=1}^{\mu} p_{hi} e_{ij} \qquad \begin{pmatrix} h = 1,2,\ldots,\mu \\ j = \mu+1,\mu+2,\ldots,m \end{pmatrix} .$$

For, on putting

$$F_i = f_i + \sum_{j=\mu+1}^{m} e_{ij} f_j \qquad (i = 1,2,\ldots,\mu) ,$$

evidently

$$\lambda_h(\mathbf{f}) = \sum_{i=1}^{\mu} p_{hi} F_i \qquad (h = 1,2,\ldots,\mu) .$$

This is a system of linear equations for $F_1, \ldots, F_\mu$ with the determinant $P \neq 0$ and can be solved in the form

(18):
$$PF_i = \sum_{h=1}^{\mu} P_{ih} \lambda_h(\mathbf{f}) \qquad (i = 1,2,\ldots,\mu) ,$$

where $P_{ih}$ is the cofactor of $p_{hi}$ in $P$, hence is a polynomial and so has a non-negative order.

Denote further by $\varepsilon$ the least common denominator of the $\mu(m-\mu)$ rational functions $e_{ij}$. It follows easily from lemma (13) that there exists a positive integer $C_3$ which depends on $Q$, but not on $c$ or on $\lambda$, such that

$$\max_{i,j} \{\partial(\varepsilon), \partial(\varepsilon e_{ij})\} \leq C_3 ;$$

we could have chosen

$$C_3 = C_0^{m^2} .$$

The linear forms

$$\varepsilon F_i = \varepsilon f_i + \sum_{j=\mu+1}^{m} (\varepsilon e_{ij}) f_j \qquad (i = 1,2,\ldots,\mu)$$

in the components $f_1, \ldots, f_m$ of $\mathbf{f}$ thus have coefficients $\varepsilon \neq 0$ and $\varepsilon e_{ij}$ which are polynomials not all zero, of degrees at most $C_3$. On the other hand, $f_1, \ldots, f_m$ are by hypothesis linearly independent over $K(z)$. Therefore

$$F_1 \neq 0, \ldots, F_\mu \neq 0 ,$$

and so, by lemma (1), which is now applied for the first time, there exists a positive integer $C_4$ which depends only on $Q$ and f , but not on $\lambda$ , such that

(19): $$\max(\mathrm{ord}_c\, F_1,\ldots,\mathrm{ord}_c\, F_\mu) \le C_4 .$$

Finally put

$$Y = \mathrm{ord}_c\, \lambda(\mathrm{f}) ;$$

since $\lambda(\mathrm{f}) \ne 0$ , $Y$ is a finite integer. From

$$\lambda_{h+1}(\mathrm{f}) = D\lambda_h(\mathrm{f}) ,$$

where as before $D$ is the operator

$$D = \kappa \frac{d}{dz} ,$$

evidently

$$\mathrm{ord}_c\, \lambda_{h+1}(\mathrm{f}) \ge \mathrm{ord}_c\, \lambda_h(\mathrm{f}) - 1$$

and therefore

(20): $$\mathrm{ord}_c\, \lambda_h(\mathrm{f}) \ge \mathrm{ord}_c\, \lambda_1(\mathrm{f}) - (h-1) = Y - (h-1) \ge Y - (\mu-1) \qquad (h = 1,2,\ldots,\mu) .$$

From (18), (19), and (20), we deduce the inequality

$$\mathrm{ord}_c\, P \ge Y - (\mu-1) - C_4 .$$

We have also obtained the upper bound (17) for $\mathrm{ord}_c\, P$ . On combining these two results and using once more that $\mu \le m - 1$ , it follows that

$$Y - (m-1)X \le Y - \mu X \le \frac{\mu(\mu-1)}{2} C_2 + C_4 + (\mu-1) \le \frac{(m-1)(m-2)}{2} C_2 + C_4 + (m-2) .$$

Put

$$C = \frac{(m-1)(m-2)}{2} C_2 + C_4 + (m-1)$$

so that $C$ depends only on $Q$ and f , but is independent of $c$ or $\lambda$ . We have then proved that

$$Y - (m-1)X < C \quad \text{if} \quad \mu \le m - 1 .$$

Conversely, if $Y - (m-1)X \ge C$ , then necessarily $\mu = m$ . Thus we arrive at the following basic result.

**THEOREM 6.** *Let*

$$Q : w_h' = \sum_{k-1}^{m} q_{hk} w_k \qquad (h = 1,2,\ldots,m)$$

*be a system of homogeneous linear differential equations with coefficients*

*$q_{hk}$ in $K(z)$. Let $c$ be an arbitrary constant, and let* f *be a solution of $Q$ which has components in $K(z-c)$ that are linearly independent over $K(z)$. Then a positive integer $C$ exists which depends only on $Q$ and* f *and has the following property.*

*Denote by $\kappa$ the least common denominator of the $m^2$ rational functions $q_{hk}$, by $p_1, \ldots, p_m$ any $m$ polynomials not all zero, and by $p_{hk}$ the polynomials defined by*

$$p_{1k} = p_k \text{ for all } k \text{, and } p_{h+1,k} = \kappa p'_{hk} + \sum_{j=1}^{m} p_{hj}(\kappa q_{jk}) \text{ for all } h \text{ and } k \text{.}$$

*The linear forms*

$$\lambda(\mathrm{w}) = p_1 w_1 + \ldots + p_m w_m \quad \text{and} \quad \lambda_h(\mathrm{w}) = p_{h1} w_1 + \ldots + p_{hm} w_m$$

*are thus connected by the identities*

$$\lambda_1(\mathrm{w}) = \lambda(\mathrm{w}), \quad \lambda_{h+1}(\mathrm{w}) = \kappa \frac{d}{dz} \lambda_h(\mathrm{w})$$

*whenever* w *is a solution of $Q$.*

*If now*

(21): $$\mathrm{ord}_c\, \lambda(\mathrm{w}) - (m-1)\max\{\partial(p_1), \ldots, \partial(p_m)\} \geq C,$$

*then $\lambda_1, \ldots, \lambda_m$ are linearly independent, and hence the determinant*

$$P = \begin{vmatrix} p_{11} & \cdots & p_{1m} \\ \vdots & & \vdots \\ p_{m1} & \cdots & p_{mm} \end{vmatrix}$$

*is distinct from zero.*

This surprisingly simple lemma forms the basis for Shidlovski's generalisation of Siegel's work on linear differential equations.

**59.** This generalisation depends on a special case of Theorem 6 and is obtained by the following considerations.

Let $\phi$ be any real number in the interval

$$0 < \phi < 1;$$

let $n_0$ be the constant

$$n_0 = \frac{C+1}{1-\phi},$$

where $C$ is defined as in Theorem 6; and let $n$ be a positive variable integer. We impose now on the polynomials $p_k$ the restrictions

(22): $$X = \max\{\partial(p_1),\ldots,\partial(p_m)\} \le n - 1 \,, \quad Y = \operatorname{ord}_c \lambda(\mathbf{f}) \ge mn - [\phi n] - 1 \,.$$

As will be seen in the next chapter, these restrictions can always be satisfied non-trivially. From (22),

$$Y - (m-1)X \ge \{mn-[\phi n]-1\} - (m-1)(n-1) = m + n - [\phi n] - 2 \ge (1-\phi)n - 1$$

and therefore

$$Y - (m-1)X \ge C \quad \text{if} \quad n \ge n_0 \,.$$

The condition (21) of Theorem 6 is then satisfied for all $n \ge n_0$, and hence also

$$P \ne 0 \quad \text{if} \quad n \ge n_0 \,.$$

In other words, as soon as $n \ge n_0$, the rank $\mu$ always has its largest possible value $m$.

Therefore, if $\mathbf{w}$ is any solution of $Q$, not necessarily with components in $K\langle z-c\rangle$, the equations

$$\lambda_h(\mathbf{w}) = \sum_{i=1}^{m} p_{hi} w_i \qquad (h = 1,2,\ldots,m)$$

can be solved for the components of $\mathbf{w}$ in the form

(23): $$P w_i = \sum_{h=1}^{m} P_{ih} \lambda_h(\mathbf{w}) \qquad (i = 1,2,\ldots,m) \,,$$

where, as before, $P_{ih}$ denotes the cofactor of $p_{hi}$ in the determinant $P$.

On substituting $\mathbf{w} = \mathbf{f}$, we find in particular that

(24): $$P f_i = \sum_{h=1}^{m} P_{ih} \lambda_h(\mathbf{f}) \qquad (i = 1,2,\ldots,m) \,.$$

Put

$$Z = \min(\operatorname{ord}_c f_1,\ldots,\operatorname{ord}_c f_m) \,.$$

On repeating the estimations in §58 for $\mu = m$ and with the formulae (18) now replaced by the new formulae (24), we find that

$$\partial(P) \le mX + \frac{m(m-1)}{2} C_2 \,, \quad \operatorname{ord}_c \lambda_h(\mathbf{f}) \ge Y - (m-1) \qquad (h = 1,2,\ldots,m) \,,$$

where $C_2$ is defined as in §57.

Therefore, by (24), also

$$\mathrm{ord}_c\, P \geq Y - (m-1) - Z \ .$$

Hence, on putting

$$\delta = \mathrm{ord}_c\, P \ , \quad \Delta = \partial(P) - \mathrm{ord}_c\, P \ ,$$

evidently

$$\Delta \leq \left\{m(n-1) + \frac{m(m-1)}{2} C_2\right\} - \{(mn-[\phi n]-1)-(m-1)-Z\} \leq [\phi n] + n_1 \ ,$$

where $n_1$ denotes the positive number

$$n_1 = \frac{m(m-1)}{2} C_2 + Z$$

which depends on the system $Q$ and on its solution **f** , but is independent of the integer $n$ .

In this notation, $P$ allows the factorisation

(25): $$P = (z-c)^{\delta}\Pi$$

where $\Pi \neq 0$ is a certain polynomial in $K[z]$ of exact degree $\Delta$ .

**60.** Denote next by $\alpha$ any element of $K$ for which

$$\kappa(\alpha) \neq 0 \ , \quad \alpha \neq c \ ,$$

so that the differential equations $Q$ are regular with respect to $K\langle z-\alpha\rangle$ . It follows then, from §48, that if

$$w_1(\alpha), \ \ldots, \ w_m(\alpha)$$

denote $m$ arbitrary elements of $K$ , then there exists a unique solution of $Q$ , **w** say, with components

$$w_k = \sum_{l=0}^{\infty} w_{k(l)}(z-\alpha)^l \qquad (k = 1,2,\ldots,m)$$

in $K\langle z-\alpha\rangle$ , say, where the constant terms are given by

$$w_{k(0)} = w_k(\alpha) \qquad (k = 1,2,\ldots,m) \ .$$

As we saw, the polynomial $\Pi \neq 0$ in (25) has the exact degree $\Delta$ . There is therefore also an integer $\nu$ such that

(26): $$0 \leq \nu \leq \Delta \ , \quad \Pi(\alpha) = \Pi'(\alpha) = \ldots = \Pi^{(\nu-1)}(\alpha) = 0 \ , \text{ but } \Pi^{(\nu)}(\alpha) \neq 0 \ .$$

With $\nu$ so defined, apply the operator

$$D = \kappa \frac{d}{dz}$$

$\nu$ times to the identities (23). Since

$$D\lambda_h(\mathrm{w}) = \lambda_{h+1}(\mathrm{w}) \quad \text{and} \quad Dw_i = \sum_{k=1}^{m} \kappa q_{ik} w_k \qquad (i = 1,2,\dots,m) ,$$

where by the definition of $\kappa$ the products $\kappa q_{ik}$ are polynomials, we obtain a new set of $m$ identities of the form

$$\text{(27):} \qquad \kappa^{\nu} P^{(\nu)} w_i + \sum_{\sigma=0}^{\nu-1} P^{(\sigma)} l_{\sigma i}(\mathrm{w}) = \sum_{k=1}^{m+\nu} p_{ik}^{*} \lambda_k(\mathrm{w}) \qquad (i = 1,2,\dots,m) .$$

Here the expressions

$$l_{\sigma i}(\mathrm{w}) = \pi_{\sigma i 1} w_1 + \dots + \pi_{\sigma i m} w_m \qquad \begin{pmatrix} \sigma = 0,1,\dots,\nu-1 \\ i = 1,2,\dots,m \end{pmatrix}$$

on the left-hand side are certain $m\nu$ linear forms in the components of $\mathrm{w}$ with coefficients in $K[z]$, and the coefficients $p_{ik}^{*}$ on the right-hand side are certain elements of the same polynomial ring $K[z]$.

We finally replace all terms in (27) by their power series in $K\langle z-\alpha\rangle$, but retain everywhere only the constant terms, that is, we apply the substitution $z = \alpha$. Then, by (26), the contribution from the sum $\sum_{\sigma=0}^{\nu-1}$ is equal to zero, and we arrive at a set of $m$ equations

$$\text{(28):} \qquad w_i(\alpha) = \{\kappa(\alpha)^{\nu} P^{(\nu)}(\alpha)\}^{-1} \sum_{h=1}^{m+\nu} p_{ih}^{*}(\alpha) \lambda_h\big(\mathrm{w}(\alpha)\big) \qquad (i = 1,2,\dots,m) ,$$

where we have put

$$\text{(29):} \qquad \lambda_h\big(\mathrm{w}(\alpha)\big) = p_{h1}(\alpha) w_1(\alpha) + \dots + p_{hm}(\alpha) w_m(\alpha) \qquad (h = 1,2,\dots,m+\nu) .$$

Here

$$\kappa(\alpha)^{\nu} P^{(\nu)}(\alpha) \neq 0$$

by the definition of $\alpha$ and $\nu$. Hence the constant terms $w_i(\alpha)$ of the components of $\mathrm{w}$ can be written as homogeneous linear expressions in the $m + \nu$ linear forms (29) with coefficients in $K$. Since these constant terms can be chosen arbitrarily in $K$, they play the role of independent indeterminates. It follows that certain $m$ of the linear forms (29), say the forms

$$\lambda_{h_1}\big(\mathrm{w}(\alpha)\big), \lambda_{h_2}\big(\mathrm{w}(\alpha)\big), \dots, \lambda_{h_m}\big(\mathrm{w}(\alpha)\big) , \text{ where } 1 \leq h_1 < h_2 < \dots < h_m \leq m + \nu ,$$

are linearly independent and thus their determinant does not vanish.

Since

$$m + \nu \le m + \Delta \le [\phi n] + m + n_1 ,$$

the result so obtained may be formulated as follows.

THEOREM 7. *Let* $K$ *be a field of characteristic zero, and let* $c$ *be an element of* $K$ *,*

$$Q : w_h' = \sum_{k=1}^{m} q_{hk} w_k \qquad (h = 1,2,\dots,m)$$

*a system of homogeneous linear differential equations with coefficients in* $K(z)$ *, and* $\mathbf{f}$ *a solution of* $Q$ *with components* $f_1, \dots, f_m$ *in* $K\langle z-c\rangle$ *that are linearly independent over* $K(z)$ *. Define* $\kappa, p_k, p_{hk},$ $\lambda(\mathbf{W})$ *, and* $\lambda_h(\mathbf{W})$ *as in Theorem 6, and denote by* $\phi$ *a constant in the interval*

$$0 < \phi < 1 ,$$

*and by* $\alpha$ *an element of* $K$ *for which*

$$\kappa(\alpha) \ne 0 , \quad \alpha \ne c .$$

*Then there exist two positive integers* $n_0$ *and* $n_1$ *with the following properties: if* $n$ *is any integer such that* $n \ge n_0$ *and*

$$\max\{\partial(p_1),\dots,\partial(p_m)\} \le n - 1 , \quad \text{ord}_c\, \lambda(\mathbf{f}) \ge mn - [\phi n] - 1 ,$$

*then* $m$ *suffixes* $h_1, h_2, \dots, h_m$ *can be selected such that*

$$1 \le h_1 < h_2 < \dots < h_m \le [\phi n] + m + n_1$$

*and that the determinant*

$$\begin{vmatrix} p_{h_1 1}(\alpha) & \dots & p_{h_1 m}(\alpha) \\ \vdots & & \vdots \\ p_{h_m 1}(\alpha) & \dots & p_{h_m m}(\alpha) \end{vmatrix}$$

*is distinct from zero.*

Under much more restrictive conditions, this theorem was first proved by C.L. Siegel (1929 and 1949) who, naturally, did not yet have available Shidlovski's powerful Theorem 6. The importance of Theorem 7 will become evident in the next chapter.

The following remark has some interest. We had found the two estimates

$$\partial(P) \le mX + \frac{m(m-1)}{2} C_1 , \quad \text{ord}_c\, P \ge Y - (m-1) - Z$$

which are valid as soon as $Y-(m-1)X$ is sufficiently large, so that $P \neq 0$. Since

$$\partial(P) \geq \text{ord}_c\, P\,,$$

this implies that

$$Y \leq mX + \frac{m(m-1)}{2}C_1 + (m-1) + Z\,,$$

which, in explicit form, is equivalent to

$$\text{ord}_c\, \lambda(\mathbf{f}) \leq m\, \max\{\partial(p_1),\ldots,\partial(p_m)\} + \min(\text{ord}_c\, f_1,\ldots,\text{ord}_c\, f_m) + \frac{m(m-1)}{2}C_2 + (m-1)\,.$$

Here $C_2$ was defined by

$$C_2 = \max\{\partial(\kappa),\partial(\kappa q_{hk})\}\,.$$

This inequality for $\text{ord}_c\, \lambda(\mathbf{f})$ remains valid under the weak hypothesis of Theorem 6. In the terminology of my paper (Mahler 1968), it expresses the fact that the components of $\mathbf{f}$ form an *almost perfect system* relative to the sequence

$$\{c,c,c,\ldots\}\,,$$

at least when the $m$ parameters $\rho_1, \rho_2, \ldots, \rho_m$ of that paper coincide.

## CHAPTER 5

## LINEAR DIFFERENTIAL EQUATIONS: A LOWER BOUND FOR THE RANK OF THE VALUES OF FINITELY MANY SIEGEL E-FUNCTIONS AT ALGEBRAIC POINTS

**61.** Let $K$ be a given finite algebraic number field. The present chapter is concerned with the linear dependence or independence over $K$ of the values of certain entire functions at points $z = \alpha$, where $\alpha \in K$. These functions are assumed to satisfy a system of homogeneous or inhomogeneous linear differential equations

$$w_h' = q_{h0} + \sum_{k=1}^{m} q_{hk} w_k \qquad (h = 1,2,\ldots,m) ,$$

where the $q_{hk}$ are rational functions in $K(z)$. The possible inhomogeneity of such a system in which some $q_{h0}$ may be $\neq 0$ will be emphasized by referring to it as a system $Q_I$.

By means of the Theorem 7 just proved we shall establish a result which in this generality was first obtained by Shidlovski (1962a), although the method of proof goes back to Siegel (1949). Unfortunately, merely because of the method of proof, really satisfactory results can at present be obtained only when $w_1 = f_1(z), \ldots, w_m = f_m(z)$ are entire functions of a very restricted kind, namely when they are *Siegel E-functions*.

Before considering such $E$-functions, it is convenient first to collect a number of properties of algebraic number fields, and to prove a theorem on systems of linear Diophantine equations, since these properties will soon be needed.

The main result proved in this chapter (Theorem 8), will be used in the next chapter to establish general theorems on the transcendency and the algebraic independence of values of Siegel $E$-functions.

**62.** As before, let $\mathbb{Q}$ and $\mathbb{C}$ be the fields of rational and complex numbers, respectively, and let $\theta$ be an arbitrary (real or complex) algebraic number, say of degree $N$ over $\mathbb{Q}$. Also let $\theta = \theta^{(0)}, \theta^{(1)}, \ldots, \theta^{(N-1)}$ be the algebraic conjugates of $\theta$ in $\mathbb{C}$. On adjoining $\theta$ to $\mathbb{Q}$, we obtain the *algebraic number field* $K = \mathbb{Q}(\theta)$ which has as its elements the rational functions

$$\alpha = r(\theta)$$

of $\theta$ with rational coefficients. The similar values

$$\alpha^{(j)} = r(\theta^{(j)}) \qquad (j = 0,1,\ldots,N-1)$$

are the *conjugates* of $\alpha$ relative to $K$.

The maximum

$$\max(|\alpha|,|\alpha^{(1)}|,\ldots,|\alpha^{(N-1)}|)\ , \ = \overline{|\alpha|} \text{ say,}$$

plays an important role in the theory of transcendental numbers because, if $\alpha$ and $\beta$ are any two elements of $K$, then

(1): $$\overline{|\alpha\mp\beta|} \le \overline{|\alpha|} + \overline{|\beta|}\ , \quad \overline{|\alpha\beta|} \le \overline{|\alpha|}\,\overline{|\beta|}\ .$$

An *algebraic integer* is defined by the property that the highest coefficient of its minimal equation is equal to 1. (Note that if $\alpha$ is a non-zero rational integer, then $\overline{|\alpha|} \ge 1$.) The set, $O_K$ say, of all algebraic integers in $K$ forms a domain of integrity of which $K$ is the quotient field. To every element $\alpha$ of $K$ there exists a smallest positive (rational) integer $a = a(\alpha)$ such that $a\alpha$ is in $O_K$; this integer $a$ is called the *denominator of* $\alpha$.

One can in many ways select a *basis* $\omega_1, \ldots, \omega_N$ of $O_K$ such that every element $\gamma$ of $O_K$ allows a unique representation

$$\gamma = g_1\omega_1 + \ldots + g_N\omega_N$$

with rational integral coefficients $g_1, \ldots, g_N$. If

$$\gamma^{(j)},\ \omega_1^{(j)},\ \ldots,\ \omega_N^{(j)} \qquad (j = 0,1,\ldots,N-1)$$

are the conjugates of $\gamma, \omega_1, \ldots, \omega_N$, respectively, then also

(2): $$\gamma^{(j)} = g_1\omega_1^{(j)} + \ldots + g_N\omega_N^{(j)} \qquad (j = 0,1,\ldots,N-1)\ ,$$

and it is known that the determinant

$$\begin{vmatrix} \omega_1^{(0)} & \cdots & \omega_N^{(0)} \\ \vdots & & \vdots \\ \omega_1^{(N-1)} & \cdots & \omega_N^{(N-1)} \end{vmatrix}$$

does not vanish. The equations (2) can then be solved for $g_1, \ldots, g_N$ in the form

(3): $$g_k = \sum_{j=0}^{N-1} \Omega_{jk}\gamma^{(j)} \qquad (k = 1,2,\ldots,N)$$

where the new coefficients $\Omega_{jk}$ depend only on the chosen basis $\omega_1, \ldots, \omega_N$ of $O_K$ and are independent of $\gamma$. From (2) and (3) it follows now that there exist two positive constants $c_1$ and $c_2$ which depend only on the ring $O_K$ and are independent of $\gamma$, such that

(4): $$\overline{|\gamma|} \le c_1 \max(|g_1|, \ldots, |g_N|) \quad \text{and} \quad \max(|g_1|, \ldots, |g_N|) \le c_2 \overline{|\gamma|} .$$

It is proved in algebra that, if we adjoin to $\mathbb{Q}$ not one, but finitely many, algebraic numbers, then the resulting extension field can also be obtained from $\mathbb{Q}$ by adjoining only a *single* suitably chosen algebraic number. On account of this fact, without loss of generality the field $K = \mathbb{Q}(\theta)$ may always be assumed to contain any given finite set of algebraic numbers.

**63.** Many proofs of transcendency are based on certain existence theorems from the theory of linear Diophantine equations and inequalities that are derived from the famous *Dirichlet's Principle* (the Schubfachprinzip) which states:

*If a set of $n+1$ elements is subdivided into $n$ subsets, then at least one of these subsets contains at least two elements.*

As a first application of this principle we prove the following theorem.

(5): *There exists a positive constant $c_0$ which depends only on $O_K$, with the following property.*

*Let*

$$y_h = \sum_{i=1}^{q} a_{hi} x_i \qquad (h = 1,2,\ldots,p)$$

*where $p < q$, be a set of $p$ linear forms in $q$ variables with coefficients $a_{hi}$ in $O_K$, and let*

$$A = \max_{h,i} \overline{|a_{hi}|} > 0 .$$

*Then integers $x_1, \ldots, x_q$ in $O_K$ not all zero exist for which*

$$\max(\overline{|x_1|}, \ldots, \overline{|x_q|}) \le c_0 (c_0 . qN . A)^{p/(q-p)} , \quad y_1 = \ldots = y_p = 0 .$$

Proof. Let the basis $\omega_1, \ldots, \omega_N$ of $O_K$ be chosen fixed once for all. Since the coefficients $a_{hi}$ are in $O_K$, so also are the products $a_{hi}\omega_j$, and hence there exist unique rational integers $a_{hijk}$ such that

$$a_{hi}\omega_j = \sum_{k=1}^{N} a_{hijk}\omega_k \qquad \begin{pmatrix} h = 1,2,\ldots,p \\ i = 1,2,\ldots,q \\ j = 1,2,\ldots,N \end{pmatrix} .$$

Similarly, for $x_1, \ldots, x_q$ in $O_K$, there are unique rational integers $x_{ij}$ such that

$$x_i = \sum_{j=1}^{N} x_{ij}\omega_j \qquad (i = 1,2,\ldots,q) .$$

These formulae imply that

$$y_h = \sum_{k=1}^{N} y_{hk}\omega_k \qquad (h = 1,2,\ldots,p) ,$$

where we have put

$$y_{hk} = \sum_{i=1}^{q} \sum_{j=1}^{N} a_{hijk}x_{ij} \qquad \binom{h = 1,2,\ldots,p}{k = 1,2,\ldots,N} .$$

Let $c_1$ and $c_2$ be as in (4), and put

$$c_3 = [c_2 \max_j |\omega_j|] + 1$$

so that $c_3$ depends only on $O_K$. The positive integer $H$ defined by

$$0 \le (qN.c_3A)^{p/(q-p)} - 1 < 2H \le (qN.c_3A)^{p/(q-p)} + 1$$

satisfies also the inequality

(6): $(2H+1)^{qN} = (2H+1)^{(q-p)N}(2H+1)^{pN} > (qN.c_3A)^{pN}.(2H+1)^{pN} \ge (2.qN.c_3A+1)^{pN}$ ,

the final inequality holding because $H$, $q$, $N$, $c_3$, and $A$ are all $\ge 1$.

In the linear forms $y_{hk}$ let each of the variables $x_{ij}$ run independently over the $2H + 1$ values

$$0, \mp 1, \ldots, \mp H .$$

Then, firstly, the $qN$-tuple $\{x_{ij}\}$ of all $x_{ij}$ has

$$(2H+1)^{qN} .$$

distinct possibilities. Next, by the second inequality (4),

$$\max_{h,i,j,k} |a_{hijk}| \le c_2 . \max_{h,i,j} \overline{|a_{hi}\omega_j|} \le c_2 . \max_{h,i} \overline{|a_{hi}|} \max_j \overline{|\omega_j|} \le c_3A ,$$

whence

$$\max_{h,k} |y_{hk}| \le q.N.c_3A.H .$$

Therefore, secondly, the $pN$-tuple $\{y_{hk}\}$ of all $y_{hk}$ has not more than

$$(2.qN.c_3A.H+1)^{pN}$$

possibilities.

Thus, by (6), there are more distinct vectors $\{x_{ij}\}$ than there are distinct vectors $\{y_{hk}\}$. This implies by Dirichlet's principle that there are two distinct vectors $\{x'_{ij}\}$ and $\{x''_{ij}\}$ to which corresponds the same vector $\{y_{hk}\}$. Put

$$x_{ij} = x'_{ij} - x''_{ij} \qquad \begin{pmatrix} i = 1,2,\ldots,q \\ j = 1,2,\ldots,N \end{pmatrix} .$$

The $qN$ integers $x_{ij}$ are then not all zero; they satisfy the inequalities

$$\max_{i,j} |x_{ij}| \leq H + H \leq (qN.c_3A)^{p/(q-p)} + 1 \ ;$$

and the $pN$ corresponding integers $y_{hk}$ all vanish simultaneously, hence the $p$ algebraic integers $y_h$ also vanish.

Finally let $c_0$ be the constant

$$c_0 = \max(2c_1, c_3) .$$

Then, from the first inequality (4),

$$\max_i \overline{|x_i|} \leq c_1 \cdot \max_{i,j} |x_{ij}| \leq c_1\{(qNc_3A)^{p/(q-p)}+1\} \leq$$

$$\leq 2c_1(qNc_3A)^{p/(q-p)} \leq c_0(c_0qNA)^{p/(q-p)} ,$$

as was to be proved.

**64.** A power series

$$f(z) = \sum_{\nu=0}^{\infty} f_\nu \frac{z^\nu}{\nu!}$$

is said to be an *E-function relative to* $K = \mathbb{Q}(\theta)$ if it satisfies the following three conditions.

(7): *All the coefficients* $f_0, f_1, f_2, \ldots$ *of* $f(z)$ *lie in* $K$.

(8): *However small the constant* $\varepsilon > 0$ *is chosen*,

$$\overline{|f_\nu|} = O(\nu^{\varepsilon\nu}) \quad as \quad \nu \to \infty .$$

(9): *Denote by* $d_\nu$ *the smallest positive integer such that the products*

$$d_\nu f_0, d_\nu f_1, \ldots, d_\nu f_\nu$$

*are in* $O_K$ . *Then, however small the constant* $\varepsilon > 0$ *is chosen,*

$$d_\nu = O(\nu^{\varepsilon\nu}) \quad \text{as} \quad \nu \to \infty .$$

Here the $O$-notation has the meaning usual in analysis and number theory. Thus $O(\nu^{\varepsilon\nu})$ denotes a function $\chi(\nu)$ (not necessarily the same function in different formulae) satisfying

$$|\chi(\nu)| \le \mu_0 \nu^{\varepsilon\nu} ,$$

where $\mu_0$ is a positive constant independent of $\nu$ , but which may depend on $\varepsilon$ .

In the following computations we shall again and again obtain estimates of the form $O(\nu^{k\varepsilon\nu})$ where $k$ is some positive constant. Such an estimate may be replaced each time by the estimate $O(\nu^{\varepsilon\nu})$ , because $\varepsilon$ may be chosen arbitrarily small and may therefore be replaced by $\varepsilon/k$ . A similar change may be made in other estimates involving a product $k\varepsilon$ .

An $E$-function will be called a *Siegel E-function* if it satisfies a linear differential equation

(10): $$w^{(n)} + r_1 w^{(n-1)} + \dots + r_n w + r_0 = 0$$

where the coefficients $r_1, \dots, r_n, r_0$ are rational functions. This distinction between $E$-functions and Siegel $E$-functions is for convenience only and has no historical justification. A second characterization of Siegel $E$-functions is contained in the following lemma.

(11): *Let* $w_1 = f_1(z), \dots, w_m = f_m(z)$ *be a solution of the system of linear differential equations*

$$Q_I : w_h' = q_{h0} + \sum_{k=1}^{m} q_{hk} w_k \qquad (h = 1,2,\dots,m)$$

*where the coefficients* $q_{hk}$ *are rational functions. If* $f_1(z)$ *is an E-function, then it is a Siegel E-function.*

*Conversely, every Siegel E-function occurs as a component of a solution of a suitable system* $Q_I$ *with rational functions as coefficients.*

Proof. By the equations $Q_I$ all the successive derivatives of $f_1(z)$ can be written as linear polynomials

$$f_1^{(h)}(z) = r_{h0}(z) + \sum_{k=1}^{m} r_{hk}(z) f_k(z) \qquad (h = 0,1,2,\dots)$$

in $f_1(z), \ldots, f_m(z)$ with coefficients $r_{hk}(z)$ that are rational functions. Since any $m + 1$ such linear polynomials are linearly dependent, $f_1(z)$ satisfies then a differential equation of the form (10), of order $n = m$.

Conversely, if $f(z)$ satisfies the differential equation (10), let

$$w_1 = f(z) , \quad w_2 = f'(z), \ldots, w_n = f^{(n-1)}(z) ;$$

then evidently

$$w_1' = w_2 , \quad w_2' = w_3, \ldots, w_{n-1}' = w_n , \quad w_n' = -(r_1 w_n + \ldots + r_n w_1 + r_0) ,$$

which is a system of differential equations of the form $Q_I$.

**65.** A number of simple properties of $E$-functions and in particular of Siegel $E$-functions are immediate consequences of their definitions.

(12): *Every polynomial in* $K[z]$ *is a Siegel* $E$*-function.*

(13): *If* $f(z)$ *is a (Siegel)* $E$*-function and* $\alpha$ *is a number in* $K$, *then* $f(\alpha z)$ *is a (Siegel)* $E$*-function.*

(14): *If* $f(z)$ *is a (Siegel)* $E$*-function, then all the derivatives*

$$f'(z),\ f''(z), \ldots$$

*are (Siegel)* $E$*-functions.*

(15): *If* $f(z)$ *is a (Siegel)* $E$*-function, then the integrals*

$$\int_0^z f(z)dz, \int_0^z \left(\int_0^z f(z)dz\right)dz, \ldots$$

*are (Siegel)* $E$*-functions.*

(16): *Every* $E$*-function is an entire function.*

There is a further important property of $E$-functions which is less obvious and requires a proof.

(17): *If* $f(z)$ *and* $g(z)$ *are (Siegel)* $E$*-functions, then*

$$f(z) + g(z) ,\quad f(z) - g(z) ,\ \textit{and}\ f(z)g(z)$$

*likewise are (Siegel)* $E$*-functions.*

Proof In explicit form, let

$$f(z) = \sum_{\nu=0}^{\infty} f_\nu \frac{z^\nu}{\nu!} \quad \text{and} \quad g(z) = \sum_{\nu=0}^{\infty} g_\nu \frac{z^\nu}{\nu!} ,$$

and denote by $e_\nu$ the positive integer belonging to $g(z)$ that is analogous to the positive integer $d_\nu$ which, by (9), belongs to $f(z)$. Then

$$f(z)+g(z)=\sum_{\nu=0}^{\infty}(f_\nu+g_\nu)\frac{z^\nu}{\nu!}\,,\quad f(z)-g(z)=\sum_{\nu=0}^{\infty}(f_\nu-g_\nu)\frac{z^\nu}{\nu!}\,,$$

$$f(z)g(z)=\sum_{\nu=0}^{\infty}h_\nu\frac{z^\nu}{\nu!}\,,$$

where $h_\nu$ is defined by

$$h_\nu=\sum_{\rho=0}^{\nu}\binom{\nu}{\rho}f_\rho g_{\nu-\rho}\,.$$

The new series obviously have the properties (7). Next,

$$\overline{|f_\nu\mp g_\nu|}\le\overline{|f_\nu|}+\overline{|g_\nu|}=O(\nu^{\varepsilon\nu})+O(\nu^{\varepsilon\nu})=O(\nu^{\varepsilon\nu})$$

and

$$\overline{|h_\nu|}\le\sum_{\rho=0}^{\nu}\binom{\nu}{\rho}\max_{0\le\rho\le\nu}\overline{|f_\rho|}\cdot\max_{0\le\rho\le\nu}\overline{|g_{\nu-\rho}|}=2^\nu O(\nu^{\varepsilon\nu})O(\nu^{\varepsilon\nu})=O(\nu^{3\varepsilon\nu})\,,$$

so that the three new series all have the property (8). They also have the property (9) because the products

$$d_\nu e_\nu(f_\rho+g_\rho)\,,\quad d_\nu e_\nu(f_\rho-g_\rho)\,,\quad d_\nu e_\nu h_\rho \qquad (\rho=1,2,\dots,\nu)$$

are in $O_K$ and since further

$$d_\nu e_\nu=O(\nu^{\varepsilon\nu})O(\nu^{\varepsilon\nu})=O(\nu^{2\varepsilon\nu})\,.$$

Consider, finally, the case when $f(z)$ and $g(z)$ are *Siegel* $E$-functions. Denote by

$$w^{(n)}+r_1w^{(n-1)}+\dots+r_nw+r_0=0 \quad\text{and}\quad w^{(p)}+s_1w^{(p-1)}+\dots+s_pw+s_0=0$$

linear differential equations with rational coefficients for $f(z)$ and $g(z)$, respectively. From these equations, each derivative of $f(z)+g(z)$ and of $f(z)-g(z)$ can be written as a linear polynomial with rational coefficients in

$$f(z),\ f'(z),\ \dots,\ f^{(n-1)}(z),\ g(z),\ g'(z),\ \dots,\ g^{(p-1)}(z)\,,$$

while each derivative of $f(z)g(z)$ is such a linear polynomial in the functions

$$f^{(h)}(z),\ g^{(k)}(z),\ f^{(h)}(z)g^{(k)}(z) \qquad \binom{h=0,1,\dots,n-1}{k=0,1,\dots,p-1}\,.$$

By a simple elimination we obtain then for $f(z)+g(z)$, $f(z)-g(z)$, and $f(z)g(z)$ linear differential equations with rational coefficients of the orders

$n+p$ , $n+p$ , and $np+n+p$ , respectively, showing that these new functions are *Siegel* $E$-functions.

**66.** We return now to the study of the solutions of linear differential equations. Let again

$$Q : w_h' = \sum_{k=1}^{m} q_{hk} w_k \qquad (h = 1,2,\ldots,m)$$

be a system of *homogeneous* linear differential equations with coefficients $q_{hk}$ in the rational function field $K(z)$ , where $K = \mathbb{Q}(\theta)$ from now on is to be an algebraic number field. We shall now only be concerned with solutions

$$\mathrm{f}(z) = \begin{bmatrix} f_1(z) \\ \vdots \\ f_m(z) \end{bmatrix} \neq 0$$

of $Q$ the components $f_1(z), \ldots, f_m(z)$ of which are $E$-functions, and hence, by (11), are Siegel $E$-functions.

We have already defined the *rank* $\rho$ *of* $\mathrm{f}(z)$ over $K(z)$ , where $z$ is an indeterminate, as the maximum number of components of $\mathrm{f}(z)$ that are linearly independent over $K(z)$ .

Similarly, if $\alpha \in K$ , let $\rho(\alpha)$ denote the rank of the constant vector

$$\mathrm{f}(\alpha) = \begin{bmatrix} f_1(\alpha) \\ \vdots \\ f_m(\alpha) \end{bmatrix}$$

*over* $K$ , that is the largest number of components $f_1(\alpha), \ldots, f_m(\alpha)$ of $\mathrm{f}(\alpha)$ that are linearly independent over $K$ . It may of course happen that $\rho(\alpha) < \rho$ .

As a basis for the investigations of the next chapter we shall show in the present chapter that, if $\alpha \neq 0$ and $\kappa(\alpha) \neq 0$ , then the quotient $\rho(\alpha)/\rho$ cannot be less than a certain positive number that depends only on the field $K$ .

**67.** Since they are $E$-functions, the components of $\mathrm{f}(z)$ can be written as power series

$$f_h(z) = \sum_{\nu=0}^{\infty} f_{h\nu} \frac{z^\nu}{\nu!} \qquad (h = 1,2,\ldots,m) ,$$

where the coefficients $f_{h\nu}$ have the properties (7), (8), and (9). Thus, by (9), to each function $f_h(z)$ and to each suffix $\nu \geq 0$ there exists a positive integer $d_{h\nu}$ such that the products

$$d_{h\nu}f_{h0},\ d_{h\nu}f_{h1},\ \ldots,\ d_{h\nu}f_{h\nu}$$

are in $O_K$ and however small $\varepsilon > 0$ is chosen,

$$d_{h\nu} = O(\nu^{\varepsilon\nu}) \quad \text{as} \quad \nu \to \infty \qquad (h = 1,2,\ldots,m) .$$

On putting

$$d_\nu = \prod_{h=1}^{m} d_{h\nu} ,$$

so that $d_\nu$ is independent of $h$, the products

$$d_\nu f_{h0},\ d_\nu f_{h1},\ \ldots,\ d_\nu f_{h\nu}$$

likewise are in $O_K$. Further

$$d_\nu = \prod_{h=1}^{m} O(\nu^{\varepsilon\nu}) = O(\nu^{m\varepsilon\nu}) ,$$

and hence by the remark in §64, also

$$d_\nu = O(\nu^{\varepsilon\nu}) .$$

68. The linear form

$$\lambda\{f(z)\} = \sum_{h=1}^{m} p_h(z)f_h(z)$$

with its derived forms

$$\lambda_h\{f(z)\} = D^{h-1}\lambda\{f(z)\} = \sum_{k=1}^{m} p_{hk}(z)f_k(z) \qquad (h = 1,2,3,\ldots)$$

have been studied in the last chapter. We shall now select the polynomials $p_h(z)$ in such a way that $\lambda$ vanishes at $z = 0$ to a very high order. As we shall see, this is possible without either the degrees or the coefficients of the $p_h(z)$ becoming too large.

For the present let $m \geq 2$, since the next considerations become trivial when $m = 1$. Denote by $n$ a positive integer, and write the polynomials $p_h(z)$ in the form

$$p_h(z) = (n-1)! \sum_{\mu=0}^{n-1} g_{h\mu} \frac{z^\mu}{\mu!} \qquad (h = 1,2,\dots,m)$$

where the coefficients $g_{h\mu}$ will soon be chosen as elements of $O_K$. The linear form $\lambda$ can then be expanded into a power series

$$\lambda\{f(z)\} = \sum_{\nu=0}^{\infty} a_\nu \frac{z^\nu}{\nu!} .$$

Here, from the power series for the components $f_h(z)$ of $f(z)$,

(18): $$a_\nu = (n-1)! \sum_{h=1}^{m} \sum_{\rho=0}^{\min(\nu,n-1)} \binom{\nu}{\rho} g_{h\rho} f_{h,\nu-\rho} \qquad (\nu = 0,1,2,\dots) .$$

Next denote by $\phi$ a second constant satisfying

$$0 < \phi < 1 .$$

In terms of this constant, we define the numbers $p$ and $q$ of Lemma (5) by

$$p = mn - [\phi n] - 1 , \quad q = mn .$$

Hence, for sufficiently large $n$,

(19): $$n \le (m-\phi)n - 1 \le p < (m-\phi)n < mn , \quad p/(q-p) = \frac{mn-[\phi n]-1}{[\phi n]+1} \le \frac{m}{\phi} .$$

The $p$ expressions

$$y_\nu = d_{p-1} \frac{a_\nu}{(n-1)!} \qquad (\nu = 0,1,\dots,p-1)$$

are linear forms in the $q$ unknowns

$$x_{h\mu} = g_{h\mu} \qquad \left(\begin{matrix} h = 1,2,\dots,m \\ \mu = 0,1,\dots,n-1 \end{matrix}\right) .$$

By the definition of $d_{p-1}$, the coefficients

$$a_{h,\nu,\rho} = d_{p-1} \binom{\nu}{\rho} f_{h,\nu-\rho} \qquad (0 \le \rho \le \nu \le p-1)$$

of these linear forms lie in $O_K$.

Also, for $0 \le \rho \le \nu \le p-1$,

$$\binom{\nu}{\rho} \le 2^\nu \le 2^{mn} , \quad d_{p-1} = O\left((mn)^{\varepsilon mn}\right) , \quad \overline{|f_{h,\nu-\rho}|} = O\left((mn)^{\varepsilon mn}\right) ,$$

$$\overline{|a_{h,\nu,\rho}|} = 2^{mn} . O\left((mn)^{\varepsilon mn}\right) . O\left((mn)^{\varepsilon mn}\right) = O\left((mn)^{3\varepsilon mn}\right) ,$$

so that, again by the remark in §64, for $n \to \infty$,

(20): $$\overline{|a_{h,\nu,\rho}|} = O(n^{\varepsilon n}) .$$

Lemma (5) can be applied to the $p$ linear forms $y_\nu$ in the $q$ unknowns $x_{h\mu}$ with the value

$$A = O(n^{\varepsilon n}) .$$

The lemma implies then that there exist $q$ integers $g_{h\mu}$ in $O_K$, not all zero, such that

(i) the $p$ linear forms $y_\nu$ vanish simultaneously, thus

(21): $$a_0 = a_1 = \dots = a_{p-1} = 0 ;$$

and

(ii) the $q$ coefficients $g_{h\mu}$ allow, by (19), the estimate

$$\max_{h,\mu} \overline{|g_{h\mu}|} \leq c_0\{c_0.mn.N.O(n^{\varepsilon n})\}^{p/(q-p)} = O(n^{2.\varepsilon n.m/\phi}) .$$

Again by the remark in §64, we are allowed to replace $2.\varepsilon.m/\phi$ by $\varepsilon$, thus finding that

$$\max_{h,\mu} \overline{|g_{h\mu}|} = O(n^{\varepsilon n}) .$$

It is convenient to introduce the new coefficients

$$G_{h\mu} = \frac{(n-1)!}{\mu!} g_{h\mu} \qquad \begin{pmatrix} h = 1,2,\dots,m \\ \mu = 0,1,\dots,n-1 \end{pmatrix} ,$$

which also lie in $O_K$ and do not all vanish. Since $(n-1)! \leq n^n$,

(22): $$\max_{h,\mu} \overline{|G_{h\mu}|} = O(n^{(1+\varepsilon)n}) .$$

The polynomials $p_h(z)$ assume now the simpler form

$$p_h(z) = \sum_{\mu=0}^{n-1} G_{h\mu} z^\mu \qquad (h = 1,2,\dots,m) .$$

Consider finally those Taylor coefficients $a_\nu$ that belong to suffixes

$$\nu \geq p .$$

By (19), for sufficiently large $n$ and for all such suffixes $\nu$, since $m \geq 2$,

$$\nu \geq p \geq n \text{ and hence } n^{\varepsilon n} = O(\nu^{\varepsilon\nu}) .$$

Therefore from (18), from

$$\sum_{\rho=0}^{\nu} \binom{\nu}{\rho} = 2^{\nu} = O(\nu^{\varepsilon\nu}) ,$$

and from the estimates given for $(n-1)!$, $\overline{|g_{h\mu}|}$, and $\overline{|f_{h\nu}|}$, it follows that

$$\overline{|a_\nu|} \leq n^n.m.O(\nu^{\varepsilon\nu})O(\nu^{\varepsilon\nu}) = n^n O(\nu^{3\varepsilon\nu}) .$$

Once more by the remark in §64, we may write $\varepsilon$ instead of $3\varepsilon$. It follows that, for $\nu \to \infty$, for all sufficiently large $n$,

(23): $$\overline{|a_\nu|} = n^n O(\nu^{\varepsilon\nu}) \quad \text{if} \quad \nu \geq p .$$

The existence result so proved is due to Siegel (1929) and Shidlovski (1966) and may be formulated as a lemma.

(24): *Let* $m \geq 2$ ; *let the components* $f_1(z), \ldots, f_m(z)$ *of* $\mathrm{f}(z)$ *be E-functions relative to* $K$ ; *let* $\varepsilon$ *and* $\phi$, *where* $0 < \varepsilon < 1$, $0 < \phi < 1$, *be constants; and let* $n$ *be a sufficiently large positive integer. Put*

$$p = mn - [\phi n] - 1 .$$

*Then there exist* $m$ *polynomials*

$$p_h(z) = \sum_{\mu=0}^{n-1} G_{h\mu} z^{\mu} \qquad (h = 1,2,\ldots,m)$$

*not all identically zero, wity coefficients* $G_{h\mu}$ *in* $O_K$ *satisfying the inequality* (22), *such that*

$$\lambda\{\mathrm{f}(z)\} = \sum_{h=1}^{m} p_h(z) f_h(z) = \sum_{\nu=p}^{\infty} a_\nu \frac{z^\nu}{\nu!} ,$$

*where the Taylor coefficients* $a_\nu$ *allow the estimate* (23).

69. From now on let $\mathrm{f}(z)$ in Lemma (24) be a solution of the system $Q$ of (11). As before, denote by $\kappa = \kappa(z)$ the least common denominator of the $m^2$ rational functions $q_{hk}$ ; here, without loss of generality, the coefficients of $\kappa$ lie in $O_K$. Therefore, by the equations

(25): $p_{1k} = p_k$ for all $k$, and $p_{h+1,k} = \kappa p'_{hk} + \sum_{j=1}^{m} p_{hj}.\kappa q_{jk}$ for all $h$ and $k$,

which determine the $p_{hk}$ in the derived forms

$$\lambda_h(w) = D^{h-1}\lambda(w) = \sum_{k=1}^{m} p_{hk}w_k \qquad (h = 1,2,3,\ldots) ,$$

also the $p_{hk}$ are polynomials with coefficients in $O_K$.

We require upper bounds for the absolute values of the conjugates relative to $K$ of the coefficients of these polynomials $p_{hk}$, and also upper bounds for the absolute values of the Taylor coefficients, $a_{h\nu}$ say, in the power series

(26): $$\lambda_h\{f(z)\} = \left(\kappa \frac{d}{dz}\right)^{h-1} \sum_{\nu=p}^{\infty} a_\nu \frac{z^\nu}{\nu!} , = \sum_{\nu=p-h+1}^{\infty} a_{h\nu} \frac{z^\nu}{\nu!} , \text{ say.}$$

The easiest way of finding such estimates is by means of majorants.

To define the majorant notation, denote by

$$u = \sum_{\nu=0}^{\infty} u_\nu z^\nu \quad \text{and} \quad v = \sum_{\nu=0}^{\infty} v_\nu z^\nu , \text{ where } v_\nu \geq 0 \text{ for all } \nu ,$$

any two formal, or convergent, power series; we do not exclude the case where these series degenerate into polynomials. If the $u_\nu$ are in $\mathbf{C}$, the notation

$$u \ll v \text{ is to mean that } |u_\nu| \leq v_\nu \text{ for all } \nu ,$$

and if the $u_\nu$ are in $K$, the similar notation

$$u \lll v \text{ is to mean that } \overline{|u_\nu|} \leq v_\nu \text{ for all } \nu .$$

Such majorant relations may evidently be added, multiplied, and differentiated.

There obviously exist two positive constants $\Gamma$ and $C$ such that

(27): $$-\kappa(z) \lll \Gamma(1+z)^C \text{ and } \kappa(z)q_{hk}(z) \lll \Gamma(1+z)^C \qquad (h,k = 1,2,\ldots,m) ;$$

here we may take for $C$ the constant

$$C_2 = \max_{h,k} \left(\partial(\kappa), \partial(\kappa q_{hk})\right)$$

introducted in §57.

We assert, firstly, that

(28): $$p_{hk}(z) \lll O(n^{(1+\varepsilon)n}).\Gamma^{h-1}(1+z)^{n-1+(h-1)C} \prod_{\nu=0}^{h-2} (\nu C+m+n-1) \text{ for all } h \text{ and } k .$$

Here the product on the right-hand side becomes empty when $h = 1$ and is then to mean 1.

In fact, when $h = 1$, the estimate (22) implies that

$$p_{1k}(z) \lll O(n^{(1+\varepsilon)n})(1+z)^{n-1} \quad \text{for all } k ,$$

so that (28) certainly is true in this case. Assume that this majorant has already been established for some suffix $h \geq 1$ ; its proof for the suffix $h+1$ and thus for all suffixes is as follows.

From (25), (27), and (28),

$$p_{h+1,k}(z) \lll \Gamma(1+z)^C.O(n^{(1+\varepsilon)n}).\Gamma^{h-1}.\left\{\prod_{\nu=0}^{h-2}(\nu C+m+n-1)\right\}.\left(m+\frac{d}{dz}\right)(1+z)^{n-1+(h-1)C}$$

$$\lll O(n^{(1+\varepsilon)n}).\Gamma^{h}(1+z)^{n-1+hC}\prod_{\nu=0}^{h-1}(\nu C+m+n-1) ,$$

whence the assertion. In this estimate we made use of the trivial formula

$$\left(m+\frac{d}{dz}\right)(1+z)^{n-1+(h-1)C} \lll \{(h-1)C+m+n-1\}(1+z)^{n-1+(h-1)C} .$$

Denote now by $\alpha$ a fixed number in $K$ , and by $n_1$ the constant integer defined in Theorem 7, and restrict the suffix $h$ to the interval

$$1 \leq h \leq [\phi n] + m + n_1 .$$

Then for $n \to \infty$ ,

$$\Gamma^{h-1}\overline{|(1+\alpha)^{n-1+(h-1)C}|} = O(n^{\varepsilon n}) , \quad \prod_{\nu=0}^{h-2}(\nu C+m+n-1) = O(n^{(\phi+\varepsilon)n}) .$$

Hence, by (28),

$$\overline{|p_{hk}(\alpha)|} = O(n^{(1+\varepsilon)n}).O(n^{\varepsilon n}).O(n^{(\phi+\varepsilon)n}) = O(n^{(1+\phi+3\varepsilon)n}) ,$$

whence, on applying once more the remark in §64,

(29): $\overline{|p_{hk}(\alpha)|} = O(n^{(1+\phi+\varepsilon)n})$ for $1 \leq h \leq [\phi n] + m + n_1$ , $k = 1, 2, \ldots, m$ .

**70.** Next denote by $L(z)$ the infinite series

$$L(z) = \sum_{\nu=p}^{\infty} |a_\nu| \frac{z^\nu}{\nu!} ,$$

so that evidently

$$\lambda_1\{f(z)\} = \lambda\{f(z)\} \ll L(z) .$$

With the same convention for empty products as in §69, we assert, secondly, that

(30): $$\lambda_h\{f(z)\} \ll \Gamma^{h-1}(1+z)^{(h-1)C}\prod_{\nu=0}^{h-2}\left(\nu C+\frac{d}{dz}\right)L(z) .$$

This assertion certainly holds for $h = 1$. Assume it has already been proved for a suffix $h \geq 1$. We shall show that it then is true also for the suffix $h + 1$ and therefore is true for all values of $h$.

In fact, on applying the operator $D$ to the majorant (30),

$$\lambda_{h+1}\{f(z)\} = \kappa \frac{d}{dz} \lambda_h\{f(z)\} << \Gamma(1+z)^C \frac{d}{dz} \left\{\Gamma^{h-1}(1+z)^{(h-1)C} \prod_{\nu=0}^{h-2} \left(\nu C + \frac{d}{dz}\right)\right\} L(z) <<$$

$$<< \Gamma(1+z)^C \{\Gamma^{h-1}.(h-1)C(1+z)^{(h-1)C-1} + \Gamma^{h-1}.(1+z)^{(h-1)C} \frac{d}{dz}\} \prod_{\nu=0}^{h-2} \left(\nu C + \frac{d}{dz}\right).L(z)$$

$$<< \Gamma^h(1+z)^{hC} \prod_{\nu=0}^{h-1} \left(\nu C + \frac{d}{dz}\right).L(z) ,$$

because trivially

$$(1+z)^{(h-1)C-1} << (1+z)^{(h-1)C} .$$

This establishes the truth of (30).

The right-hand side of (30) can be estimated as follows. Evidently

$$\prod_{\nu=0}^{h-2} \left(\nu C + \frac{d}{dz}\right)L(z) << \left\{\prod_{\nu=0}^{h-2} (\nu C)\right\}\left(1 + \frac{d}{dz}\right)^{h-1} L(z) << (hC)^h \sum_{\eta=0}^{h} \binom{h}{\eta} \frac{d^\eta L(z)}{dz^\eta} .$$

Here, for $0 \leq \eta \leq h$, by (23),

$$\frac{d^\eta L(z)}{dz^\eta} = n^n \frac{d^\eta}{dz^\eta} \sum_{\nu=p}^{\infty} O(\nu^{\varepsilon\nu}) . \frac{z^\nu}{\nu!} << n^n \sum_{\nu=p-h}^{\infty} O\left((\nu+h)^{\varepsilon(\nu+h)}\right) \frac{z^\nu}{\nu!} .$$

Since

$$\sum_{\eta} \binom{h}{\eta} = 2^h ,$$

the majorant (30) implies then that

$$(31): \qquad \lambda_h\{f(z)\} << \Gamma^{h-1}(1+z)^{(h-1)C}(2hC)^h n^n \sum_{\nu=p-h}^{\infty} O\left((\nu+h)^{\varepsilon(\nu+h)}\right) \frac{z^\nu}{\nu!} .$$

Here the integer $p$ was defined by

$$p = mn - [\phi n] - 1 ,$$

while the suffix $h$ is as before restricted to the interval

$$1 \leq h \leq [\phi n] + m + n_1 .$$

From now on, let $\phi$ satisfy the stronger inequality

$$0 < \phi < \min(1,m/5) .$$

Then for all sufficiently large $n$,

$$p - h \ge mn - 2[\phi n] - m - n_1 - 1 \ge (m-3\phi)n \ge 2\phi n \ge h .$$

The sum $\sum_{\nu=p-h}^{\infty}$ extends therefore only over suffixes $\nu \ge h$ , and for these suffixes

$$(\nu+h)^{\varepsilon(\nu+h)} \le (2\nu)^{2\varepsilon\nu} .$$

Since

$$\nu! \ge \nu^{\nu}e^{-\nu} ,$$

it follows that for sufficiently large $n$ ,

$$\sum_{\nu=p-h}^{\infty} (\nu+h)^{\varepsilon(\nu+h)} \frac{|\alpha|^{\nu}}{\nu!} \le \sum_{\nu\ge(m-3\phi)n} \frac{(2\nu)^{2\varepsilon\nu}|\alpha|^{\nu}e^{\nu}}{\nu^{\nu}} \le \sum_{\nu\ge(m-3\phi)n} \frac{(2^{2\varepsilon}e|\alpha|)^{\nu}}{\nu^{(1-2\varepsilon)\nu}}$$

and that therefore

(32): $$\sum_{\nu=p-h}^{\infty} O\left((\nu+h)^{\varepsilon(\nu+h)}\right) \frac{|\alpha|^{\nu}}{\nu!} = O(n^{-(1-3\varepsilon)(m-3\phi)n}) .$$

Thus, on applying (31) with $z = \alpha$ , we find that for $n \to \infty$ ,

$$|\lambda_h\{f(\alpha)\}| \le \Gamma^{2\phi n}(1+|\alpha|)^{2\phi Cn}(4\phi Cn)^{2\phi n}n^n O(n^{-(1-3\varepsilon)(m-3\phi)n}) =$$

$$= O(n^{(1+3\phi)n-(1-3\varepsilon)(m-3\phi)n}) .$$

Here assume that $\varepsilon$ is already so small that

$$(1-3\varepsilon)(m-3\phi) \ge m - 4\phi .$$

It follows then finally that for $n \to \infty$ ,

(33): $$|\lambda_h\{f(\alpha)\}| = O(n^{-(m-1-7\phi)n}) \quad \text{if} \quad 1 \le h \le [\phi n] + m + n_1 .$$

The coefficient 7 of $\phi$ has no importance since $\phi$ may be chosen arbitrarily small. We repeat that this estimate has been established only for $m \ge 2$ .

**71.** Suppose for the moment that the algebraic number $\alpha$ in $K$ satisfies

$$\alpha \ne 0 \text{ , but } \kappa(\alpha) = 0 .$$

We assert that then the $m$ function values

$$f_1(\alpha), \ldots, f_m(\alpha)$$

are *linearly dependent over* $K$ . For the differential equations $Q$ imply that

$$0 = \kappa(\alpha)f_h'(\alpha) = \sum_{k=1}^{m} \lim_{z\to\alpha} \{\kappa(z)q_{hk}(z)\}.f_k(\alpha) \qquad (h = 1,2,\ldots,m) ,$$

and here, by the definition of $\kappa$ , at least one of the coefficients

$$\lim_{z\to\alpha}\{\kappa(z)q_{hk}(z)\} \qquad (h,k = 1,2,\ldots,m)$$

is distinct from zero. All these coefficients lie in $K$, whence the assertion.

This trivial case will from now on be excluded, and it will henceforth be assumed that

(34): $$\alpha \in K, \quad \alpha \neq 0, \quad \kappa(\alpha) \neq 0 .$$

*It is still unknown whether this hypothesis implies that the* $m$ *function values* $f_1(\alpha), \ldots, f_m(\alpha)$ *are linearly independent over* $K$. As we shall see, this independence property does, however, hold if $K$ either is the rational number field $\mathbf{Q}$ or is any imaginary quadratic number field.

When $K$ is a general algebraic number field, we can prove only a weaker result; however, this weaker result will suffice for proofs of transcendency and algebraic independence in the next chapter. Namely, if as in §66 we denote by $\rho$ the rank of $\mathbf{f}(z)$ over $K(z)$ and by $\rho(\alpha)$, $= r$ say, the rank of $\mathbf{f}(\alpha)$ over $K$, then it is possible to obtain a lower bound for $\rho(\alpha)$. This is the objective of the remainder of this chapter. It is clear that $r \leq \rho$. For the present assume that

(35): $$m \geq 2, \quad \rho = m, \quad 1 \leq r \leq m-1 .$$

Here the restrictions on $m$ and on $\rho$ will later be removed. The case $r = m$ is trivial, and the case $r = 0$ cannot hold. For by the last relation in (34), $z = \alpha$ is a regular point of the system $Q$, and hence $\mathbf{f}(\alpha)$ cannot be the zero vector because this would imply that also $\mathbf{f}(z) \equiv \mathbf{0}$, contrary to the second relation in (35).

Thus exactly $r$, but no more, of the components $f_1(\alpha), \ldots, f_m(\alpha)$ of $\mathbf{f}(\alpha)$ are linearly independent over $K$. Hence there exists an $(m-r) \times m$ matrix

$$\begin{vmatrix} s_{11} & \cdots & s_{1m} \\ \vdots & & \vdots \\ s_{m-r,1} & \cdots & s_{m-r,m} \end{vmatrix}$$

of exact rank $m - r$, with elements that, without loss of generality, lie in $O_K$, such that

(36): $$s_{h1}f_1(\alpha) + \ldots + s_{hm}f_m(\alpha) = 0 \qquad (h = 1,2,\ldots,m-r) .$$

The left-hand sides of these equations provide $m - r$ linearly independent linear forms in $f_1(\alpha), \ldots, f_m(\alpha)$ with coefficients in $O_K$.

Further linear forms in these function values are given by

$$\lambda_h\{\mathbf{f}(\alpha)\} = p_{h1}(\alpha)f_1(\alpha) + \ldots + p_{hm}(\alpha)f_m(\alpha) \qquad (h = 1,2,3,\ldots)$$

where, as before, the $\lambda_h\{f(z)\}$ are derived from the form

$$\lambda\{f(z)\} = p_1(z)f_1(z) + \ldots + p_m(z)f_m(z)$$

constructed in Lemma (24). As we found, the polynomials $p_k$ have at most the degree $n-1$ ; $\lambda\{f(z)\}$ has a zero at $z = 0$ of order at least

$$p = mn - [\phi n] - 1 ;$$

and, by hypothesis, the components $f_1(z), \ldots, f_m(z)$ of $f(z)$ are entire functions that are linearly independent over $K(z)$ .

**72.** It follows that the hypothesis of Theorem 7 in Chapter 4 is satisfied. Therefore, as soon as

$$n \geq n_0 ,$$

there exist $m$ suffixes $h_1, h_2, \ldots, h_m$ satisfying

$$1 \leq h_1 < h_2 < \ldots < h_m \leq [\phi n] + m + n_1$$

such that the corresponding linear forms

$$\lambda_h\{f(\alpha)\} , \text{ where } h = h_1, h_2, \ldots, h_m ,$$

are linearly independent. Among these $m$ forms there are further $r = \rho(\alpha)$ forms

(37): $$\lambda_h\{f(\alpha)\} , \text{ where } h = j_1, j_2, \ldots, j_r ,$$

say, which are linearly independent of the $m - r$ forms

$$s_{h1}f_1(\alpha) + \ldots + s_{hm}f_m(\alpha) \qquad (h = 1,2,\ldots,m-r)$$

in the equations (36). Therefore the determinant of these $r + (m-r) = m$ forms,

$$S = \begin{vmatrix} p_{j_1,1}(\alpha) & \ldots & p_{j_1,m}(\alpha) \\ \vdots & & \vdots \\ p_{j_r,1}(\alpha) & \ldots & p_{j_r,m}(\alpha) \\ s_{1,1} & \ldots & s_{1,m} \\ \vdots & & \vdots \\ s_{m-r,1} & \ldots & s_{m-r,m} \end{vmatrix} ,$$

satisfies the inequality

(38): $$S \neq 0 .$$

Denote by $S_{ik}$ the cofactor of the element of $S$ in the $i$th row and the $k$th column. A system of $m$ linear equations for the $m$ unknowns $f_1(\alpha), \ldots, f_m(\alpha)$ is obtained by combining the $m - r$ equations (36) with the $r$ equations

$$\lambda_h\{\mathbf{f}(\alpha)\} = p_{h1}(\alpha)f_1(\alpha) + \ldots + p_{hm}(\alpha)f_m(\alpha) \text{ , where } h = j_1, j_2, \ldots, j_r .$$

By (38), this system can be solved in the form

$$(39): \qquad S.f_k(\alpha) = \sum_{i=1}^{r} S_{ik}\lambda_{j_i}\{\mathbf{f}(\alpha)\} \qquad (k = 1,2,\ldots,m) .$$

We have already remarked that $\mathbf{f}(\alpha)$ cannot be the zero vector. It is therefore possible to choose a suffix $k$ such that

$$(40): \qquad f_k(\alpha) \neq 0 .$$

73. The elements $p_{j_i,k}(\alpha)$ in the first $r$ rows of the determinant $S$ are polynomials in $\alpha$ with coefficients in $O_K$ ; as functions of $n$ , they allow the estimates (29). The elements $s_{hk}$ in the remaining $m - r$ rows of $S$ also lie in $O_K$ , but are independent of $n$ . We obtain therefore the estimate

$$(41): \qquad \overline{|S|} = O(n^{(1+\phi+\varepsilon)nr}) .$$

Similarly, for $1 \leq i \leq r$ and for $1 \leq k \leq m$ ,

$$(42): \qquad \overline{|S_{ik}|} = O(n^{(1+\phi+\varepsilon)n(r-1)}) .$$

From its definition as a determinant, $S$ is a polynomial in $\alpha$ with coefficients in $O_K$ . Here, by (25), each polynomial $p_{hk}(z)$ has at most the degree

$$(n-1) + (h-1)C ,$$

where $h$ assumes only values satisfying

$$h \leq [\phi n] + m + n_1 .$$

Therefore the degree of $S$ as polynomial in $\alpha$ cannot exceed

$$\{(1+\phi C)n+O(1)\}r ,$$

which for large $n$ is not greater than

$$[(1+2\phi C)nr] .$$

Hence, if $a$ denotes the denominator of $\alpha$ , that is, the smallest positive integer such that $a\alpha$ is an algebraic integer, then the product

$$T = a^{[(1+2\phi C)nr]}S$$

is an element of $O_K$ .

Let

$$S^{(0)} = S,\ S^{(1)},\ \ldots,\ S^{(N-1)} \quad \text{and} \quad T^{(0)} = T,\ T^{(1)},\ \ldots,\ T^{(N-1)}$$

be the conjugates relative to $K$ of $S$ and of $T$ , respectively. By (38), all these conjugates are distinct from zero. Furthermore, all the $T^{(j)}$ are integers in $O_K$ , and

$$T^{(j)} = a^{[(1+2\phi C)nr]}S^{(j)} \qquad (j = 0,1,\ldots,N-1) .$$

The norm

$$\text{norm } T = TT^{(1)} \ldots T^{(N-1)}$$

is then a rational integer distinct from zero, so that

$$|TT^{(1)}\ldots T^{(N-1)}| \geq 1 .$$

This inequality again is equivalent to

(43): $$|S|^{-1} \leq a^{[(1+2\phi C)nr]N}|S^{(1)}\ldots S^{(N-1)}| .$$

Here, substitute for each conjugate $S^{(1)}, \ldots, S^{(N-1)}$ of $S$ its upper bound (41). We find then that

$$|S|^{-1} = e^{O(n)}\{O(n^{(1+\phi+\epsilon)nr})\}^{N-1} ,$$

which can be simplified to

(44): $$|S|^{-1} = O(n^{(1+2\phi)nr(N-1)})$$

by choosing $\epsilon$ sufficiently small in terms of $\phi$ .

In one case this estimate can be improved, namely when $K$ is a nonreal field, for then $S$ and one of its conjugates relative to $K$ , say the conjugate $S^{(N-1)}$ , are complex conjugate numbers, and therefore

$$|S| = |S^{(N-1)}| .$$

It follows thus from (43) that now

$$|S|^{-2} \leq a^{[(1+2\phi C)nr]N}|S^{(1)}\ldots S^{(N-2)}| .$$

Instead of (44) the estimate for $|S|^{-1}$ takes therefore now the form

(45): $$|S|^{-1} = O(n^{(1+2\phi)nr(N-2)/2}) .$$

The desired lower bound for $r$ will now be obtained by also finding an estimate for $|S|$ in the opposite direction. For this purpose we use the equation (39) for a suffix $k$ satisfying (40), and further apply the estimates (33) and (42). We find that

$$|S| = |f_k(\alpha)|^{-1}.r.O(n^{(1+\phi+\varepsilon)n(r-1)}).O(n^{-(m-1-7\phi)n}) .$$

Depending now on whether $K$ is a real or an imaginary field, we combine this formula with either (44) or (45). On multiplying the two estimates, we find in the first case that

$$1 = O(n^{(1+2\phi)nr(N-1)}).O(n^{(1+\phi+\varepsilon)n(r-1)}).O(n^{-(m-1-7\phi)n})$$

and in the second case that

$$1 = O(n^{(1+2\phi)nr(N-2)/2}).O(n^{(1+\phi+\varepsilon)n(r-1)}).O(n^{-(m-1-7\phi)n}) .$$

Finally, we allow $n$ to tend to infinity. It follows then at once that in the first case

$$(1+2\phi)r(N-1) + (1+\phi+\varepsilon)(r-1) - (m-1-7\phi) \geq 0 ,$$

and in the second case

$$(1+2\phi)r\left(\frac{N}{2} - 1\right) + (1+\phi+\varepsilon)(r-1) - (m-1-7\phi) \geq 0 .$$

In these inequalities, $m$, $r$, and $N$ are fixed positive integers, while $\phi$ and $\varepsilon$ may be chosen as arbitrarily small positive numbers. It follows in the first case that

$$r(N-1) + (r-1) - (m-1) = rN - m \geq 0 ,$$

and in the second case that

$$r\left(\frac{N}{2} - 1\right) + (r-1) - (m-1) = \frac{rN}{2} - m \geq 0 .$$

These estimates imply the following result.

**(46):** *Suppose that* $K$ *is an algebraic number field of degree* $N$ *, and that*

$$m \geq 2 , \quad \alpha \in K , \quad \alpha \neq 0 , \quad \kappa(\alpha) \neq 0 , \quad \rho = m .$$

*Then*

$$\rho(\alpha) \geq \begin{cases} \frac{m}{N} & \text{if } K \text{ is real,} \\ \frac{2m}{N} & \text{if } K \text{ is nonreal.} \end{cases}$$

COROLLARY. *Since* $\rho(\alpha) \leq \rho$ *, this lemma implies that*

$$\rho(\alpha) = \rho = m$$

*if* $K$ *is either the rational number field* $(N = 1)$ *, or an imaginary* *quadratic number field* $(N = 2)$ .

Both the lemma and its corollary remain valid in the excluded case when $m = 1$ . For then trivially $\frac{m}{N} \le 1$ , and $\frac{2m}{N} \le 1$ , respectively. On the other hand, $\mathbf{f}(\alpha) \ne \mathbf{0}$ and therefore $\rho(\alpha) = 1$ .

**74.** We owe to Shidlovski (1962) the fundamental result that Lemma (46) can be generalised to the case when $\rho < m$ . His result may be formulated as follows.

**THEOREM 8.** *Let* $K$ *be an algebraic number field of degree* $N$ *over* $\mathbf{Q}$ . *Let*

$$Q : w_h' = \sum_{k=1}^{m} q_{hk} w_k \qquad (h = 1,2,\ldots,m)$$

*be a system of homogeneous linear differential equations with coefficients* $q_{hk}$ *in* $K(z)$ , *and let* $\kappa$ *be the least common denominator of these coefficients. Let* $\mathbf{f}(z) \not\equiv \mathbf{0}$ *be a solution of* $Q$ *such that its components* $f_1(z), \ldots, f_m(z)$ *are E-functions relative to* $K$ . *Assume that exactly* $\rho$ , *but no more, of these components are linearly independent over* $K(z)$ . *Let* $\alpha$ *be an algebraic number in* $K$ *satisfying*

$$\alpha \ne 0 , \quad \kappa(\alpha) \ne 0 ,$$

*and suppose that exactly* $\rho(\alpha)$ , *but no more, of the function values* $f_1(\alpha), \ldots, f_m(\alpha)$ *are linearly independent over* $K$ .

*Then*

$$\rho(\alpha) \ge \frac{\rho}{N} \quad \text{if } K \text{ is real,}$$

$$\rho(\alpha) \ge \frac{2\rho}{N} \quad \text{if } K \text{ is nonreal.}$$

**COROLLARY.** *Since again* $\rho(\alpha) \le \rho$ , *it follows that* $\rho(\alpha) = \rho$ *when* $K$ *is either the rational number field or an imaginary quadratic number field.*

**Proof of the theorem.** The hypothesis $\kappa(\alpha) \ne 0$ implies that $z = \alpha$ is a regular point of the system $Q$ ; that is, the components of every solution of $Q$ are regular relative to $K(z-\alpha)$ . It is then possible, by Theorem 5 of Chapter 4, to select $\rho$ components $f_{h_1}(z), \ldots, f_{h_\rho}(z)$ of $\mathbf{f}(z)$ with the following two properties.

(i): These $\rho$ components are linearly independent over $K(z)$ .

(ii): They satisfy a new system $Q^0$ of linear differential equations with coefficients $q_{hk}^0$ in $K(z)$ which are likewise regular at $z = \alpha$ .

It suffices now to apply Lemma (46) to the $\rho$-vector $\mathbf{f}^0(z)$ with the components $f_{h_1}(z), \ldots, f_{h_\rho}(z)$, so that $m$ is replaced by $\rho$, and the assertion follows immediately.

When $K$ is neither the rational field nor an imaginary quadratic field, Theorem 8 does not imply that $\rho(\alpha) = \rho$, but provides only a weaker result. It would be of interest to decide whether the lower bounds $\frac{\rho}{N}$ and $\frac{2\rho}{N}$ for $\rho(\alpha)$ given by this theorem are still best-possible for general number fields $K$.

**75.** To conclude this chapter, we deduce from Theorem 8 a consequence for systems of inhomogeneous linear differential equations.

(47): *Let*

$$Q^* : w_h' = q_{h0} + \sum_{k=1}^{m} q_{hk} w_k \qquad (h = 1,2,\ldots,m)$$

*be a system of inhomogeneous linear differential equations with coefficients* $q_{h0}, q_{hk}$ *in* $K(z)$, *and let* $\kappa$ *be the least common denominator of these coefficients. Let* $w_1 = f_1(z), \ldots, w_m = f_m(z)$ *be a solution of* $Q^*$ *where* $f_1(z), \ldots, f_m(z)$ *are E-functions relative to* $K$. *Let* $\alpha$ *be a number satisfying*

$$\alpha \in K, \quad \alpha \neq 0, \quad \kappa(\alpha) \neq 0 .$$

*Assume further that* $\rho^*$, *but no more, of the functions* $f_1(z), \ldots, f_m(z), 1$ *are linearly independent over* $K(z)$, *and that* $\rho^*(\alpha)$, *but no more, of the function values* $f_1(\alpha), \ldots, f_m(\alpha), 1$ *are linearly independent over* $K$. *Then*

$$\rho^*(\alpha) \geq \frac{\rho^*}{N} \quad \textit{if } K \textit{ is a real number field,}$$

$$\rho^*(\alpha) \geq \frac{2\rho^*}{N} \quad \textit{if } K \textit{ is a nonreal number field.}$$

**Proof.** In addition to $f_1(z), \ldots, f_m(z)$ introduce the further $E$-function

$$f_0(z) \equiv 1 .$$

Then the $(m+1)$-vector with the components $w_0 = f_0(z), w_1 = f_1(z), \ldots, w_m = f_m(z)$ forms a solution of the following homogeneous system of linear differential equations:

$$Q : w_0' = 0 \ , \quad w_h' = q_{h0}w_0 + \sum_{k=1}^{m} q_{hk}w_k \qquad (h = 1,2,\ldots,m)$$

The assertion is thus an immediate consequence of Theorem 8.

# CHAPTER 6

## LINEAR DIFFERENTIAL EQUATIONS: SHIDLOVSKI'S THEOREMS ON THE TRANSCENDENCY AND ALGEBRAIC INDEPENDENCE OF VALUES OF SIEGEL E-FUNCTIONS

**76.** Let

$$f(z) = \sum_{\nu=0}^{\infty} f_\nu \frac{z^\nu}{\nu!}$$

be a transcendental $E$-function relative to some algebraic number field $K$. If $\alpha \neq 0$ is any algebraic number, it is in general *not* possible to be certain whether $f(\alpha)$, or more generally $f^{(k)}(\alpha)$, is an *algebraic* or a *transcendental* number. This is evident from the following two examples.

Firstly, choose

$$f_\nu = \begin{cases} 1 & \text{if } \nu = n! \\ 0 & \text{for all other } \nu \end{cases} \qquad (n = 1,2,3,\ldots) .$$

Then $f(z)$ is a transcendental $E$-function relative to $\mathbf{Q}$. From Theorem 3 of Chapter 1 it is easy to deduce that $f^{(k)}(\alpha)$ is transcendental for all algebraic $\alpha \neq 0$ and for all $k \geq 0$.

Secondly, let

$$f(z) = \sum_{r=1}^{\infty} \frac{z^{t_{r-1}} B_r(z)}{t_{r-1}!}$$

where the notation is the same as in §36, and where $t_{r-1}$ again tends sufficiently rapidly to infinity. The same proof as in §36 shows that $f(z)$ and all its derivatives are algebraic at all algebraic points, and it is also easy to prove that $f(z)$ is transcendental and an $E$-function relative to $\mathbf{Q}$.

On the other hand, if we assume that $f(z)$ more specially is a *Siegel* $E$-function, the position changes completely: it then becomes possible to deduce from Theorem 8 of the preceding chapter that $f(z)$, and more generally any one of its derivatives, can be algebraic at only *finitely many* algebraic points $z = \alpha$. In fact, Shidlovski (1962) established even more general results on the *algebraic independence* of the values of any finite number of Siegel $E$-functions at a given algebraic point $z = \alpha$. These results form a particularly valuable part of

Shidlovski's work.

The proofs of these theorems are of an algebraic nature, once Theorem 8 of the preceding chapter has been obtained. We shall base them on certain simple estimates for the number of linearly independent polynomials of a given degree that vanish at a given point $(x_1,\dots,x_n)$ . In a classical paper, Hilbert (1890) determined an exact formula for this number, and he did so for the much more general problem of the polynomials in a polynomial ideal. For our purpose, the weaker result suffices, and it has the advantage of having a short and elementary proof.

The method used by Shidlovski himself is rather different, and is less simple.

**77.** Let $L$ be a field of characteristic zero which contains the field $Q$ of all rational numbers as a subfield and is itself contained in some larger field $L^*$ . A polynomial

$$P(w_1,\dots,w_n) = \sum c_{h_1\dots h_n} w_1^{h_1} \dots w_n^{h_n} \not\equiv 0$$

in $L[w_1,\dots,w_n]$ has the *total degree*

$$\nabla(P) = \max(h_1+\dots+h_n)$$

where the maximum extends over all systems of suffixes $h_1, \dots, h_n$ for which

(1): $$c_{h_1\dots h_n} \neq 0 .$$

When $P$ is a *homogeneous* polynomial or, as we say, an $H$-polynomial,

$$\nabla(P) = h_1 + \dots + h_n$$

for all terms of $P$ satisfying (1).

Let $x_1, \dots, x_n$ be finitely many elements of $L^*$ . These elements are called *algebraically dependent* over $L$ , or *algebraically independent* over $L$ , respectively, according as there exists, or does not exist, a polynomial $P(w_1,\dots,w_n)$ in $L[w_1,\dots,w_n]$ satisfying both

(2): $$P(w_1,\dots,w_n) \not\equiv 0 \quad \text{and} \quad P(x_1,\dots,x_n) = 0 .$$

Similarly, $x_1, \dots, x_n$ are said to be *algebraically H-dependent*, or *algebraically H-independent*, over $L$ , respectively, according as there does, or does not, exist an $H$-polynomial in $L[w_1,\dots,w_n]$ satisfying (2).

If $P(w_1,\dots,w_{n-1})$ is any polynomial in $L[w_1,\dots,w_{n-1}]$ , say of total degree $p = \nabla(P)$ , then

$$P^*(w_1,\dots,w_n) = w_1^p P\left(\frac{w_2}{w_1}, \dots, \frac{w_n}{w_1}\right)$$

is an $H$-polynomial in $L[w_1,\dots,w_n]$ , and then again, conversely,

$$P(w_1,\dots,w_{n-1}) = P^*(1,w_1,\dots,w_{n-1}) .$$

It follows that if $x_1 \neq 0$ , and if

(3): $$y_k = \frac{x_{k+1}}{x_1} \qquad (k = 1,2,\dots,n-1) ,$$

then $x_1, \dots, x_n$ *are algebraically H-dependent (H-independent) over* $L$ *if and only if* $y_1, \dots, y_{n-1}$ *are algebraically dependent (independent) over* $L$ .

This statement can be generalised. Denote by

$$d = d_L(x_1,\dots,x_n)$$

the largest number of elements $x_1, \dots, x_n$ that are algebraically independent over $L$ . Let similarly $d_H$ be the largest number of these elements that are algebraically $H$-independent over $L$ , and put

$$D = D_L(x_1,\dots,x_n) = d_H - 1 .$$

Then, if $x_1 \neq 0$ , and if $y_1, \dots, y_{n-1}$ are defined by (3),

(4): $$D_L(x_1,\dots,x_n) = d_L(y_1,\dots,y_{n-1}) .$$

For if, say, $x_1, \dots, x_{D+1}$ are algebraically $H$-independent, then $y_1, \dots, y_D$ are algebraically independent, and vice versa.

Finally, we note the following fact regarding $H$-independence: *if* $x_1, \dots, x_n$ *are algebraically H-independent over* $L$ *, then some one of* $x_1, \dots, x_n$ *can be replaced by* $1$ *and the new system is still H-independent over* $L$ . This assertion is trivial for $n = 1$ , so suppose $n \geq 2$ . If the statement were false, then for some $H$-independent system $x_1, \dots, x_n$ , each of the systems

$$\{1,x_2,\dots,x_n\} ,$$
$$\{x_1,1,\dots,x_n\} ,$$
$$\dots$$
$$\{x_1,x_2,\dots,x_{n-1},1\}$$

would be $H$-dependent over $L$ . Hence we would have $n$ homogeneous equations over $L$ , each connecting the elements of one of the above systems; the left hand sides of course are (possibly) inhomogeneous polynomials in the $x_i$'s actually present. With no loss in generality each may be taken to be irreducible over $L$, and in this case some two of them, at least, are different and therefore relatively prime, since different $x_i$'s are missing from the various equations. Suppose that the identities connecting the first two systems above are different, and that they are

$$\sum a^{(1)}_{i_2\dots i_n} x_2^{i_2} x_3^{i_3} \dots x_n^{i_n} = 0 \,, \quad \sum a^{(2)}_{i_1 i_3\dots i_n} x_1^{i_1} x_3^{i_3} \dots x_n^{i_n} = 0 \,.$$

Rewrite these relations in the form

$$\sum a^{(1)}_{i_2\dots i_n} \left(\frac{x_2}{x_n}\right)^{i_2} \dots \left(\frac{x_{n-1}}{x_n}\right)^{i_{n-1}} x_n^{i_2+\dots+i_n} = 0 \,,$$

$$\sum a^{(2)}_{i_1 i_3\dots i_n} \left(\frac{x_1}{x_n}\right)^{i_1} \left(\frac{x_3}{x_n}\right)^{i_3} \dots \left(\frac{x_{n-1}}{x_n}\right)^{i_{n-1}} x_n^{i_1+i_3+\dots+i_n} = 0 \,.$$

Eliminating $x_n$ from these equations, we obtain a relation of the form

$$\sum b_{i_1 i_2\dots i_{n-1}} \left(\frac{x_1}{x_n}\right)^{i_1} \dots \left(\frac{x_{n-1}}{x_n}\right)^{i_{n-1}} = 0 \,,$$

where not all the $b$'s are zero. Multiplication by a suitable power $x_n^N$ gives

$$\sum b_{i_1\dots i_{n-1}} x_1^{i_1} \dots x_{n-1}^{i_{n-1}} x_n^{N-(i_1+\dots+i_{n-1})} = 0 \,,$$

a contradiction.

78. From now on let $n \geq 2$. Further, without loss of generality, let the notation be such that $x_1 \neq 0$, and such that $x_1, \dots, x_{D+1}$ are algebraically $H$-independent over $L$, but that $x_1, \dots, x_{D+1}$, $x_{k+1}$ are algebraically $H$-dependent for all suffixes $k$ satisfying $D+1 \leq k \leq n-1$. This is equivalent to the supposition that $y_1, \dots, y_D$ are algebraically independent over $L$, while $y_1, \dots, y_D$, $y_k$ are algebraically dependent over $L$ if $D+1 \leq k \leq n-1$.

Form the two extension fields

$$U = L(y_1,\dots,y_D) \quad \text{and} \quad V = L(y_1,\dots,y_{n-1}) = U(y_{D+1},\dots,y_{n-1}) \,.$$

If $D = 0$, $U$ is identical with $L$; otherwise $U$ is a purely transcendental extension of $L$. Further, if $D = n-1$, then $U$ and $V$ coincide, while otherwise $V$ is obtained by adjoining to $U$ the $n - D - 1$ elements

$y_{D+1}, \ldots, y_{n-1}$ , all of which are algebraic over $U$ .

It follows by Abel's well-known theorem on repeated adjunctions that $V$ can be defined as an extension field

(5): $$V = U(y)$$

of $U$ , where $y$ is a suitably chosen single element of $V$ which is algebraic over $U$ and which may be taken equal to 0 if $D = n - 1$ . Since we may multiply $y$ by any element of $U$ distinct from 0 without affecting the relation (5), we are allowed to assume that the minimal equation of $y$ over $U$ has the form

(6): $$A(y;y_1,\ldots,y_D) = 0$$

where

(7): $$A(w;w_1,\ldots,w_D) = w^a + \sum_{h=1}^{a} A_h(w_1,\ldots,w_D)w^{a-h}$$

is an irreducible polynomial in $L[w,w_1,\ldots,w_D]$ which is monic relative to the indeterminate $w$ .

Thus the field $V$ is algebraic over $U$ , of degree $a$ , and

$$1, y, y^2, \ldots, y^{a-1}$$

is a basis of $V$ over $U$ . The elements $y_{D+1}, \ldots, y_{n-1}$ , which lie in $V$ , can therefore be written in the form

(8): $$y_k = Y_k(y;y_1,\ldots,y_D) \qquad (k = D+1,D+2,\ldots,n-1) ,$$

where the expressions

(9): $$Y_k(w;w_1,\ldots,w_D) = \frac{\sum_{h=0}^{a-1} B_{hk}(w_1,\ldots,w_D)w^h}{B_0(w_1,\ldots,w_D)} \qquad (k = D+1,D+2,\ldots,n-1)$$

are rational functions in $L(w,w_1,\ldots,w_D)$ with the following property: the denominator

$$B_0(w_1,\ldots,w_D) \not\equiv 0$$

and all the factors $B_{hk}(w_1,\ldots,w_D)$ in the numerators are polynomials in the ring $L[w_1,\ldots,w_D]$ which are *relatively prime*. Since $y_1, \ldots, y_D$ are algebraically independent, it follows in particular that

(10): $$B_0(y_1,\ldots,y_D) \neq 0 ,$$

so that the formulae (8) are meaningful.

**79.** By the formulae (6) and (7), there holds for every suffix $l \geq 0$ an equation of the form

$$y^l = \sum_{h=0}^{a-1} C_{hl}(y_1,\ldots,y_D)y^h \qquad (l = 0,1,2,\ldots) \tag{11}$$

where the $C_{hl}(w_1,\ldots,w_D)$ are polynomials in $L[w_1,\ldots,w_D]$. For this is obvious from (6) if $0 \leq l \leq a$. If further there is a relation (11) for any exponent $l \geq a$, then there is also one for the next higher exponent, since evidently

$$\begin{aligned}(12):\quad y^{l+1} &= \sum_{h=0}^{a-1} C_{hl}(y_1,\ldots,y_D)y^{h+1} = \\ &= \sum_{h=0}^{a-2} C_{hl}(y_1,\ldots,y_D)y^{h+1} - C_{a-1,l}(y_1,\ldots,y_D)\sum_{h=1}^{a} A_h(y_1,\ldots,y_D)y^{a-h},\end{aligned}$$

which is an equation of the required form.

Denote by $t$ a positive integer and by $h_1, \ldots, h_{n-1}$ any set of $n - 1$ integers satisfying

$$h_1 \geq 0, \ldots, h_{n-1} \geq 0, h_1 + \ldots + h_{n-1} \leq t,$$

and put

$$z_{h_1\ldots h_{n-1}} = B_0(y_1,\ldots,y_D)^t y_1^{h_1} \ldots y_{n-1}^{h_{n-1}}.$$

If $D = n - 1$, this means that

$$z_{h_1\ldots h_{n-1}} = B_0(y_1,\ldots,y_D)^t y_1^{h_1} \ldots y_D^{h_D}. \tag{13}$$

Otherwise, on putting

$$h^* = h_{D+1} + \ldots + h_{n-1},$$

it follows from (8) and (9) that

$$z_{h_1\ldots h_{n-1}} = B_0(y_1,\ldots,y_D)^{t-h^*} y_1^{h_1} \ldots y_D^{h_D} \prod_{k=D+1}^{n-1} \left\{\sum_{h=0}^{a-1} B_{hk}(y_1,\ldots,y_D)y^h\right\}^{h_k}. \tag{14}$$

On multiplying out the product, this becomes

$$z_{h_1\ldots h_{n-1}} = B_0(y_1,\ldots,y_D)^{t-h^*} y_1^{h_1} \ldots y_D^{h_D} \sum_{l=0}^{(a-1)h^*} D_{lh_1\ldots h_{n-1}}(y_1,\ldots,y_D)y^l \tag{15}$$

where the $D$'s with subscripts are sums of products of $h^*$ factors $B_{hk}$. On substituting here for $y^l$ its value (11), we obtain the representation

(16): $$z_{h_1\cdots h_{n-1}} = \sum_{h=0}^{a-1} E_{hh_1\cdots h_{n-1}}(y_1,\ldots,y_D)y^h ,$$

where the $E$'s are defined by

(17): $$E_{hh_1\cdots h_{n-1}}(y_1,\ldots,y_D) =$$
$$= B_0(y_1,\ldots,y_D)^{t-h^*} y_1^{h_1} \cdots y_D^{h_D} \sum_{l=0}^{(a-1)h^*} D_{lh_1\cdots h_{n-1}}(y_1,\ldots,y_D)C_{hl}(y_1,\ldots,y_D) .$$

These formulae show that the $E$'s are polynomials in $y_1, \ldots, y_D$ with coefficients in $L$, but that these $E$'s do not involve $y_{D+1}, \ldots, y_{n-1}$.

The equations (17) take a simpler form when all the polynomials $A_h$ vanish identically, a case which happens if and only if $y = 0$ and therefore $a = 1$. Now

$$C_{hl} = \begin{cases} 1 & \text{if } h = l = 0 , \\ 0 & \text{otherwise,} \end{cases}$$

and it follows that

(18): $$E_{hh_1\cdots h_{n-1}} = \begin{cases} B_0(y_1,\ldots,y_D)^{t-h^*} D_{0h_1\cdots h_{n-1}}(y_1,\ldots,y_D)y_1^{h_1} \cdots y_D^{h_D} & \text{if } h = 0 , \\ 0 & \text{otherwise.} \end{cases}$$

**80.** We next determine upper estimates for the total degrees of the $E$'s.

So far we have defined in §77 the total degree $\nabla$ only for polynomials that do not vanish identically. If we adopt the convention that

$$\nabla(0) = -\infty ,$$

then $\nabla$ evidently has the properties

(19): $$\nabla(P \mp P^*) \le \max\big(\nabla(P),\nabla(P^*)\big) , \quad \nabla(PP^*) = \nabla(P) + \nabla(P^*) .$$

Now put

$$c_1 = \max_h \nabla(A_h) , \quad c_2(l) = \max_h \nabla(C_{hl}) , \quad c_3 = \max_{h,k} \big(\nabla(B_0),\nabla(B_{hk})\big) ,$$

$$c_4 = \max_{l,h_1,\ldots,h_{n-1}} \nabla\big(D_{lh_1\cdots h_{n-1}}\big) , \quad c_5 = \max_{h,h_1,\ldots,h_{n-1}} \nabla\big(E_{hh_1\cdots h_{n-1}}\big) .$$

Exclude for the present the case when $y = 0$ ; then $c_1 \ge 0$.

From (12) and (19),

$$c_2(l+1) \le c_2(l) + c_1 ,$$

and it is also obvious from the definition of the $C$'s that

$$c_2(0) = c_2(1) = \dots = c_2(a-1) = 0 .$$

Hence

$$c_2(l) \le c_1(l-a+1) \quad \text{for} \quad l \ge a ,$$

and thus, since $c_1 \ge 0$ ,

(20): $$c_2(l) \le c_1(a-1)h^* \quad \text{for} \quad 0 \le l \le (a-1)h^* .$$

Next, by (14), (15), and (19),

$$c_4 \le c_3 h^* .$$

Therefore, by (17), (19), and (20),

$$c_5 \le c_3(t-h^*) + (h_1+\dots+h_D) + \{c_3h^*+c_1(a-1)h^*\} .$$

Since $h_1 + \dots + h_{n-1} \le t$ , it follows then that there exists a positive integer $c$ independent of $t$ such that

(21): $$\max_{h,h_1,\dots,h_{n-1}} \nabla\left(E_{hh_1\dots h_{n-1}}\right) \le ct .$$

This inequality has been proved under the hypothesis that $y \ne 0$ . If, however, $y = 0$ , then from (18),

$$c_5 \le c_3(t-h^*) + (h_1+\dots+h_D) + c_3h^* ,$$

and hence the inequality (21) remains valid also in this case.

81. Consider now the set, $V(t)$ say, of all $H$-polynomials

(22): $$P_H(w_1,\dots,w_n) = \sum P^{(H)}_{g_1\dots g_n} w_1^{g_1} \dots w_n^{g_n}$$

in $L[w_1,\dots,w_n]$ which have the total degree $t$ ; here the summation is extended over all sets of integers $g_1, \dots, g_n$ satisfying

$$g_1 \ge 0, \dots, g_n \ge 0, \; g_1 + \dots + g_n = t ,$$

and the coefficients $P^{(H)}_{g_1\dots g_n}$ may be arbitrary elements of $L$ . Evidently $V(t)$ is a linear vector space over $L$ which, by induction on $n$ , can be easily shown to have the dimension

$$v(t) = \binom{n+t-1}{n-1}$$

over $L$ .

Next denote by $S(t)$ the set of all $H$-polynomials $P_H$ in $V(t)$ which satisfy the equation

$$P_H(x_1,\dots,x_n) = 0 ,$$

where $x_1, \dots, x_n$ are the same elements of $L^*$ as before. $S(t)$ also forms a linear vector space over $L$ and is therefore a subspace of $V(t)$ .

The $H$-polynomial $P_H$ belongs to $S(t)$ if and only if its coefficients $p^{(H)}_{g_1\dots g_n}$ satisfy a certain system, $\Sigma(t)$ say, of homogeneous linear equations with coefficients in $L$ . Denote by $h(t)$ the number of linearly independent equations of this system $\Sigma(t)$ .

The number $h(t)$ so defined is the Hilbert function of $S(t)$ . In his paper of 1890 already referred to, Hilbert proved a general theorem which establishes an explicit formula for this function. If $D = D_L(x_1,\dots,x_n)$ is defined as in §77, Hilbert's theorem asserts that

$$\text{(23):} \qquad h(t) = h_0\binom{t}{D} + h_1\binom{t}{D-1} + \dots + h_D \quad \text{for} \quad t \geq t_0 .$$

Here $h_0 > 0$ , $h_1, \dots, h_D$ , $t_0$ are certain integers that do not depend on $t$ . In other words, if $t$ is sufficiently large, then the Hilbert function $h(t)$ is an integral-valued polynomial in $t$ , of exact degree $D$ .

By way of example, let $D = n - 1$ , so that $x_1, \dots, x_n$ are algebraically $H$-independent over $L$ . This implies that $S(t)$ consists only of the single polynomial 0 and that therefore

$$h(t) = v(t) = \binom{n+t-1}{n-1} = \binom{t+D}{D} .$$

This representation of $h(t)$ has the asserted form (23).

A proof of Hilbert's theorem in its full generality requires a detailed study of polynomial ideals and is quite lengthy. Therefore we shall establish here a weaker and more special result which, however, suffices for our purpose and which we shall easily deduce from what already has been proved in §§78-80.

**82.** $V(t)$ was defined as the set of all $H$-polynomials

$$P_H(w_1,\dots,w_n) = \sum p^{(H)}_{g_1\dots g_n} w_1^{g_1} \dots w_n^{g_n}$$

in $L[w_1,\dots,w_n]$ of total degree $t$ , and $S(t)$ was then the subset of all those

$P_H$ in $V(t)$ for which

$$P_H(x_1,\dots,x_n) = 0 .$$

Again let $y_1, \dots, y_{n-1}$ be derived from $x_1, \dots, x_n$ by the relations (3). In analogy to $V(t)$ and $S(t)$, we form the set $V^*(t)$ of all (homogeneous or inhomogeneous) polynomials

$$P(w_1,\dots,w_{n-1}) = \sum_{(h)} P_{h_1\dots h_{n-1}} w_1^{h_1} \dots w_{n-1}^{h_{n-1}}$$

in $L[w_1,\dots,w_{n-1}]$ of total degree *at most* $t$, and the subset $S^*(t)$ of all those $P$ in $V^*(t)$ for which

$$P(y_1,\dots,y_{n-1}) = 0 ;$$

here the summation $\sum_{(h)}$ extends over all sets of $n - 1$ suffixes $h_1, \dots, h_{n-1}$ satisfying

$$h_1 \geq 0, \dots, h_{n-1} \geq 0, h_1 + \dots + h_{n-1} \leq t .$$

The two complementary formulae

$$P(w_1,\dots,w_{n-1}) = P_H(1,w_1,\dots,w_{n-1})$$

and

$$P_H(w_1,\dots,w_n) = w_1^t . P\left(\frac{w_2}{w_1}, \dots, \frac{w_n}{w_1}\right)$$

establish a one-to-one mapping of $V(t)$ onto $V^*(t)$ and also of $S(t)$ onto $S^*(t)$. In terms of the coefficients, this mapping has the explicit form

(24): $P_{h_1\dots h_{n-1}} = P^{(H)}_{g_1\dots g_n}$

where $g_1 = t - (h_1+\dots+h_{n-1})$, $g_2 = h_1, \dots, g_n = h_{n-1}$.

In particular, $P_H$ and $P$ have the same number of coefficients, namely, the number

$$v(t) = \binom{n+t-1}{n-1} .$$

If the polynomial $P$ in $V^*(t)$ is to belong to $S^*(t)$, its coefficients $P_{h_1\dots h_{n-1}}$ again have to satisfy a certain system of homogeneous linear equations with coefficients in $L$, the system $\Sigma^*(t)$ say. It is immediately clear from the mapping formulae (24) that the coefficient matrix of the linear equations in $\Sigma^*(t)$ is identical with the coefficient matrix of the linear equations in $\Sigma(t)$. The number $h(t)$ of linearly independent linear equations is therefore the same for

$\Sigma(t)$ and $\Sigma^*(t)$ .

The formulae in §80 allow us to determine the linear equations of $\Sigma^*(t)$ in explicit form. We had found the representation

$$B_0(y_1,\ldots,y_D)^t y_1^{h_1} \ldots y_{n-1}^{h_{n-1}} = \sum_{h=0}^{a-1} E_{hh_1\ldots h_{n-1}}(y_1,\ldots,y_D)y^h ,$$

whence it follows that

$$B_0(y_1,\ldots,y_D)^t P(y_1,\ldots,y_{n-1}) = \sum_{(h)} \sum_{h=0}^{a-1} P_{h_1\ldots h_{n-1}} E_{hh_1\ldots h_{n-1}}(y_1,\ldots,y_D)y^h .$$

Here $y_1, \ldots, y_D$ are algebraically independent over $L$ , and $y$ is algebraic of degree $a$ over the field $U = L(y_1,\ldots,y_D)$ . Since $B_0(y_1,\ldots,y_D) \neq 0$ , $P$ belongs to the set $S^*(t)$ , that is, it has the property $P(y_1,\ldots,y_{n-1}) = 0$ , if and only if the system of equations

$$\sum_{(h)} P_{h_1\ldots h_{n-1}} E_{hh_1\ldots h_{n-1}}(y_1,\ldots,y_D) = 0 \qquad (h = 0,1,\ldots,a-1)$$

is satisfied, and this is the case if and only if the identities

(25): $$\sum_{(h)} P_{h_1\ldots h_{n-1}} E_{hh_1\ldots h_{n-1}}(w_1,\ldots,w_D) \equiv 0 \qquad (h = 0,1,\ldots,a-1)$$

hold, where $w_1, \ldots, w_D$ are independent indeterminates.

The factors $E_{hh_1\ldots h_{n-1}}(w_1,\ldots,w_D)$ are polynomials in $L[w_1,\ldots,w_D]$ the total degrees of which, by (21), do not exceed the value $ct$ . It follows that each such polynomial consists of not more than

$$\binom{ct+D}{D}$$

separate monomials. On equating to zero the coefficients of all these monomials in the identities (25), we obtain therefore not more than

$$a.\binom{ct+D}{D}$$

homogeneous linear equations with coefficients in $L$ for the coefficients

$$P_{h_1\ldots h_{n-1}}$$

of $P$ , and $\Sigma^*(t)$ consists exactly of these linear equations. The number $h(t)$ of *linearly independent* linear equations in $\Sigma^*(t)$ necessarily then satisfies the inequality

(26):
$$h(t) \le a\binom{ct+D}{D} .$$

It is even simpler to obtain a *lower* estimate for $h(t)$ by means of the following consideration.

In the special case when the coefficients of $P \in V^*(t)$ have the property that

(27): $P_{h_1 \dots h_{n-1}} = 0$ whenever $h_{D+1}, \dots, h_{n-1}$ are not all equal to $0$ ,

$P$ reduces to a polynomial in the $D$ indeterminates $w_1, \dots, w_D$ and has therefore only

$$\binom{t+D}{D}$$

coefficients $P_{h_1 \dots h_D 0 \dots 0}$ . If now $P$ belongs in addition to $S^*(t)$ , so that it vanishes at $w_1 = y_1, \dots, w_D = y_D$ , then it is identically zero, because $y_1, \dots, y_D$ are algebraically independent over $L$ .

This implies that, under the hypothesis (27), the set $\Sigma^*(t)$ cannot contain fewer than $\binom{t+D}{D}$ linearly independent linear equations. On the other hand, the number of linearly independent linear equations in $\Sigma^*(t)$ cannot increase when the conditions (27) are imposed on $P$ . It follows then that always

(28):
$$h(t) \ge \binom{t+D}{D} .$$

The two estimates (26) and (28) together contain the

**(29): MAIN LEMMA.** *Let $L$ be a field which contains the rational field $\mathbb{Q}$ and is itself contained in a field $L^*$ . Let $n \ge 2$ , and let $x_1, \dots, x_n$ be $n$ elements of $L^*$ which are not all equal to $0$ . Denote by $D+1$ the largest number of these elements that are algebraically $H$-independent over $L$ , by $t$ a positive integer, by $V(t)$ the set of all $H$-polynomials $P_H$ in $L[w_1,\dots,w_n]$ which have the total degree $t$ , and by $S(t)$ the subset of all those $P_H$ in $V(t)$ which have the property $P_H(x_1,\dots,x_n) = 0$ . The polynomial $P_H \in V(t)$ belongs to $S(t)$ if and only if its coefficients satisfy a certain system $\Sigma(t)$ of homogeneous linear equations with coefficients in $L$ . Denote by $h(t)$ the maximum number of linear equations in $\Sigma(t)$ that are linearly independent.*

*Then two positive integral constants $a$ and $c$ exist, depending only on $L$ and $x_1, \dots, x_n$ , such that*

$$\binom{t+D}{D} \le h(t) \le a\binom{ct+D}{D} .$$

**83.** Two important theorems of Shidlovski now follow easily from the Main Lemma. Again let

$$Q : w_h' = \sum_{k=1}^{m} q_{hk} w_k \qquad (h = 1,2,\dots,m)$$

be a system of homogeneous linear differential equations with coefficients $q_{hk}$ in $K(z)$, where, as before, $K$ is an algebraic number field of degree $N$ over $\mathbb{Q}$. Also as before, let $\kappa$ be the least common denominator of the rational functions $q_{hk}$; let $\mathbf{f}(z)$ be a solution of $Q$ with components $f_1(z), \dots, f_m(z)$ that are $E$-functions relative to $K$; and let $\alpha$ be a number satisfying

(A): $$\alpha \in K , \quad \alpha \neq 0 , \quad \kappa(\alpha) \neq 0 .$$

In order to exclude a trivial case, it will be assumed for the present that

$$m \geq 2 .$$

Just as in the last chapter, we consider simultaneously the *variable* vector $\mathbf{f}(z)$ and the *constant* vector $\mathbf{f}(\alpha)$. The components of these two vectors will play the role of the elements $x_1, \dots, x_n$ in the Main Lemma.

Firstly apply this lemma with $n = m$, $L = K(z)$, and $L^* = L(f_1(z),\dots,f_m(z))$, and denote by $D(z) + 1 = D_{K(z)}(f_1(z),\dots,f_m(z)) + 1$ the maximum number of *functions* $f_1(z), \dots, f_m(z)$ that are algebraically $H$-independent over $K(z)$. Denote further by $S_z(t)$ the set of all $H$-polynomials $P(w_1,\dots,w_m)$ of total degree $t > 0$, with coefficients in $K(z)$, for which

(30): $$P(f_1(z),\dots,f_m(z)) \equiv 0$$

identically in $z$. In order that $P$ should have this property, its coefficients must satisfy a certain system $\Sigma_z(t)$ of homogeneous linear equations with coefficients in $K(z)$. If $h_z(t)$ is the maximum number of linear equations in $\Sigma_z(t)$ that are linearly independent, then by the Main Lemma there exist two positive integral constants $a(z)$ and $c(z)$ such that

(31): $$\binom{t+D(z)}{D(z)} \le h_z(t) \le a(z)\binom{c(z)t+D(z)}{D(z)} .$$

Secondly, apply the main lemma with $n = m$, $L = K$, and $L^* = K(f_1(\alpha),\dots,f_m(\alpha))$, and denote now by $D(\alpha) + 1 = D_K(f_1(\alpha),\dots,f_m(\alpha)) + 1$ the maximum number of *function values* $f_1(\alpha), \dots, f_m(\alpha)$ that are algebraically

$H$-independent over $K$. Let $S_\alpha(t)$ be the set of all $H$-polynomials $P(w_1,\ldots,w_m)$ of total degree $t$, with coefficients in $K$, for which

(32): $$P\big(f_1(\alpha),\ldots,f_m(\alpha)\big) = 0 .$$

In order that $P$ should have this property, its coefficients must now satisfy a system $\Sigma_\alpha(t)$ of homogeneous linear equations with coefficients in $K$. If $h_\alpha(t)$ denotes the maximum number of linear equations in $\Sigma_\alpha(t)$ that are linearly independent, it follows again from the Main Lemma that there are two positive integral constants $a(\alpha)$ and $c(\alpha)$ such that

(33): $$\binom{t+D(\alpha)}{D(\alpha)} \le h_\alpha(t) \le a(\alpha)\binom{c(\alpha)t+D(\alpha)}{D(\alpha)} .$$

84. Denote by $T(t)$ the set of all systems of $m$ integers $(h) = (h_1,\ldots,h_m)$ satisfying

$$h_1 \ge 0, \ \ldots, \ h_m \ge 0, \ h_1 + \ldots + h_m = t .$$

As we saw in §81, $T(t)$ has

$$\binom{t+m-1}{m-1}, \ = \tau \ \text{say},$$

elements $(h)$. With each such $(h)$ associate the two products

(34): $$W_{(h)} = W_{h_1\ldots h_m} = w_1^{h_1} \ldots w_m^{h_m}$$

and

(35): $$F_{(h)}(z) = F_{h_1\ldots h_m}(z) = f_1(z)^{h_1} \ldots f_m(z)^{h_m} .$$

By imposing an arbitrary, but fixed, order on the sets $(h)$ in $T(t)$, the $\tau$ products (34) become the components of a vector $\mathsf{W}$ and the $\tau$ products (35) those of a vector $\mathsf{F}(z)$. We shall also consider the constant vector $\mathsf{F}(\alpha)$ obtained from $\mathsf{F}(z)$ by the substitution $z = \alpha$.

Each $H$-polynomial $P(w_1,\ldots,w_m)$ in either $S_z(t)$ or $S_\alpha(t)$ is a linear form in the $\tau$ products (34), and this linear form has coefficients in $K(z)$, or in $K$, satisfying the homogeneous linear equations in $\Sigma_z(t)$, or in $\Sigma_\alpha(t)$, respectively.

The original vector $\mathsf{f}(t)$ was a solution of $Q$. These differential equations for $\mathsf{w}$ imply for $\mathsf{W}$ that

$$W'_{(h)} = \left(\sum_{j=1}^{m} h_j w_j^{-1} w_j'\right) W_{(h)} = \left(\sum_{j=1}^{m} h_j w_j^{-1} \sum_{k=1}^{m} q_{jk} w_k\right) W_{(h)} .$$

Hence $W$ satisfies a new system of homogeneous linear differential equations

$$Q(t) : W'_{(h)} = \sum_{(k)\in T(t)} q_{(h)(k)} W_{(k)} \quad \text{where} \quad (h) \in T(t) ,$$

and where the new coefficients $q_{(h)(k)}$ are defined for $(h) \in T(t)$ and $(k) \in T(t)$ and evidently are linear forms in the rational functions $q_{hk}$, with numerical coefficients which are rational integers. Therefore all the products $\kappa q_{(h)(k)}$ are polynomials in $K[z]$. This means in particular that if $\alpha$ has the property (A) then none of the coefficients $q_{(h)(k)}$ has a pole at $z = \alpha$.

From its definition, $\mathsf{F}(z)$ is a special solution of $Q(t)$, with components $F_{(h)}(z)$ which are again $E$-functions relative to $K$.

85. Using a notation similar to that in Theorem 8 of the last chapter, we denote by $\rho(t;z)$ the maximum number of components $F_{(h)}(z)$ of $\mathsf{F}(z)$ that are linearly independent over $K(z)$, and by $\rho(t;\alpha)$ the maximum number of components $F_{(h)}(\alpha)$ of $\mathsf{F}(\alpha)$ that are linearly independent over $K$. By Theorem 8, applied to the system $Q(t)$ and to the special solution $\mathsf{F}(z)$ of this system, the inequality

(36): $$\rho(t;\alpha) \geq \frac{\rho(t;z)}{N}$$

holds where $N$ is again the degree of $K$ over $\mathbb{Q}$.

We assert now that

(37): $$\rho(t;z) = h_z(t) \quad \text{and} \quad \rho(t;\alpha) = h_\alpha(t) .$$

Since both these assertions are proved in the same manner, it will suffice to establish only the first equation (37).

The $\tau$ components $F_{(h)}(z)$ of $\mathsf{F}(z)$ satisfy exactly those homogeneous linear equations with coefficients in $K(z)$ that are obtained from the identities

$$P\{f_1(z),\ldots,f_m(z)\} \equiv 0 ,$$

where $P$ runs over the $H$-polynomials in $S_z(t)$, if these $H$-polynomials are written as linear forms in the products (35). Now the $\tau$ coefficients of every $H$-polynomial $P$ in $S_z(t)$ have to satisfy exactly $h_z(t)$ linearly independent homogeneous linear equations with coefficients in $K(z)$; thus $S_z(t)$ contains $\tau - h_z(t)$ and no more $H$-polynomials that are linearly independent over $K(z)$. In other words, the $\tau$ components $F_{(h)}(z)$ of $\mathsf{F}(z)$ are subject to just $\tau - h_z(t)$ linearly independent homogeneous linear equations in $K(z)$. This again implies that exactly $h_z(z)$ and no more of the components of $\mathsf{F}(z)$ are linearly independent over $K(z)$ , which proves the assertion.

The inequality (36) is thus equivalent to

$$h_\alpha(t) \geq \frac{h_z(t)}{N} .$$

Here, by (31) and (33),

$$h_z(t) \geq \binom{t+D(z)}{D(z)} \quad \text{and} \quad h_\alpha(t) \leq a(\alpha)\binom{c(\alpha)t+D(\alpha)}{D(\alpha)} ,$$

so that necessarily

(38): $$a(\alpha)\binom{c(\alpha)t+D(\alpha)}{D(\alpha)} \geq \binom{t+D(z)}{D(z)} N^{-1}$$

We now allow $t$ to tend to infinity in this inequality. The left-hand side is a polynomial in $t$ of the exact degree $D(\alpha)$ , the right-hand side one of the exact degree $D(z)$ , and in both polynomials the coefficient of the highest power of $t$ is positive. It follows then from (38) that necessarily

(39): $$D(\alpha) \geq D(z) .$$

Assume for the moment that even the stronger relation

$$D(\alpha) > D(z)$$

holds. Possibly after a suitable renumbering of the suffixes, we may then also assume that the $D(\alpha) + 1$ function values

(40): $$f_1(\alpha), \ldots, f_{D(\alpha)+1}(\alpha)$$

are algebraically $H$-independent over $K$ . On the other hand, the $D(\alpha)+1 > D(z)+1$ functions

$$f_1(z), \ldots, f_{D(\alpha)+1}(z)$$

certainly are algebraically $H$-dependent over $K(z)$ .

Hence there exists an $H$-polynomial $P(w_1,\ldots,w_{D(\alpha)+1}) \not\equiv 0$ with coefficients in $K(z)$ such that

(41): $$P\left(f_1(z),\ldots,f_{D(\alpha)+1}(z)\right) \equiv 0$$

identically in $z$ . If $p \not\equiv 0$ is any rational function in $K(z)$ , the new $H$-polynomial $pP$ has the same property. Choose $p$ in such a way that all the coefficients of $pP$ become polynomials in $K[z]$ which are relatively prime, and put

$$pP = P^*(z|w_1,\ldots,w_{D(\alpha)+1}) .$$

Also $P^*$ is an $H$-polynomial not identically zero with the property (41), and if in it we put $z = \alpha$ , then all its coefficients remain finite and they do not all vanish simultaneously. The equation

$$P^*\left(\alpha \mid f_1(\alpha),\ldots,f_{D(\alpha)+1}(\alpha)\right) = 0 ,$$

which is obtained for $z = \alpha$ , has therefore finite coefficients not all zero. It thus implies that $f_1(\alpha), \ldots, f_{D(\alpha)+1}(\alpha)$ are algebraically $H$-dependent over $K$ , contrary to the hypothesis.

The assumption that $D(\alpha) > D(z)$ is therefore false, and it follows from (39) that $D(\alpha) = D(z)$ . We have thus established the following result.

(42): *Suppose that* $K$ *is a number field of finite degree over* $\mathbb{Q}$ *, that*

$$Q : w_h' = \sum_{k=1}^{m} q_{hk}w_k \qquad (h = 1,2,\ldots,m)$$

*is a system of* $m \geq 2$ *homogeneous linear equations with coefficients* $q_{hk}$ *in* $K(z)$ *, and that* $\mathrm{f}(z)$ *is a solution of* $Q$ *with components* $f_1(z), \ldots, f_m(z)$ *that are E-functions relative to* $K$ *. Let* $\alpha \neq 0$ *be a number in* $K$ *which is not a pole of any of the rational functions* $q_{hk}$ *.*

*Then the largest number of functions* $f_1(z), \ldots, f_m(z)$ *that are algebraically* $H$*-independent over* $K(z)$ *is equal to the largest number of function values* $f_1(\alpha), \ldots, f_m(\alpha)$ *that are algebraically* $H$*-independent over* $K$ *.*

**86.** The lemma just proved can be simplified, and it is in particular possible to remove from it the reference to the arbitrary number field $K$ .

Denote for the present by $K, K^*, K_1, \ldots, K_m$ any algebraic number fields which are all of finite degrees over the rational field $\mathbb{Q}$ .

If $K \subset K^*$ , if $K^*$ is of degree $n$ over $K$ , and if $\alpha$ is any element of $K$ , then the algebraic conjugates of $\alpha$ relative to $K^*$ are the same as those relative to $K$ , except that each conjugate is now counted $n$ times. The maximum $\overline{|\alpha|}$ thus depends only on $\alpha$ and not on the field of which it is considered as an element. Furthermore, $\alpha$ is or is not an algebraic integer regardless of the field $\alpha$ is considered as belonging to. From these two facts it follows immediately that *if* $f(z)$ *is an E-function relative to* $K$ *, it is also an E-function relative to any extension field* $K^*$ *.*

Next, if $K_1, \ldots, K_m$ are any finite number of algebraic number fields, there always exists an algebraic number field $K$ which contains them all as subfields. *Hence, if* $f_1(z), \ldots, f_m(z)$ *are E-functions relative to the equal or distinct fields* $K_1, \ldots, K_m$ *, respectively, then they are also E-functions relative to some algebraic number field* $K$ *which is the same for all of them.* This field $K$ can moreover be chosen so as to contain any given algebraic number $\alpha$ .

87. The system $Q$ of differential equations in Lemma (42) had its coefficients in $K(z)$ , where $K$ was an algebraic number field. The following lemma shows that this restriction on the coefficients may be relaxed.

**(43):** *Let the components* $f_1(z), \ldots, f_m(z)$ *of* $\mathrm{f}(z)$ *be E-functions relative to an algebraic number field* $K$ *; let* $\mathrm{f}(z)$ *be a solution of the system of differential equations*

$$Q : w_h' = \sum_{k=1}^{m} q_{hk} w_k \qquad (h = 1,2,\ldots,m)$$

*with coefficients* $q_{hk}$ *in* $\mathrm{C}(z)$ *; and let* $\alpha \in K$ *be distinct from the poles of these coefficients* $q_{hk}$ *. Then* $\mathrm{f}(z)$ *also satisfies a system of differential equations*

$$Q_1 : w_h' = \sum_{k=1}^{m} q_{1,hk} w_k \qquad (h = 1,2,\ldots,m)$$

*with coefficients* $q_{1,hk}$ *in* $K(z)$ *, where* $\alpha$ *is again distinct from the poles of these new coefficients* $q_{1,hk}$ *.*

Proof. Let the least common denominator $\kappa$ of the coefficients $q_{hk}$ be fixed so as to be monic in $z$ . On putting

$$\kappa_{hk} = \kappa q_{hk} \qquad (h,k = 1,2,\ldots,m) ,$$

the $m^2 + 1$ polynomials $\kappa$ and $\kappa_{hk}$ have coefficients in $\mathrm{C}$ , and the total number of these coefficients is finite. Hence there exist

(i) finitely many numbers $\chi_1, \ldots, \chi_n$ in $\mathrm{C}$ which are linearly independent over $K$ , and

(ii) a set of $(m^2+1)n$ polynomials

$$\kappa_\nu \text{ and } \kappa_{\nu,hk} \qquad (h,k = 1,2,\ldots,m;\ \nu = 1,2,\ldots,n)$$

in $K[z]$ , such that

$$\kappa = \sum_{\nu=1}^{n} \chi_\nu \kappa_\nu \text{ and } \kappa_{hk} = \sum_{\nu=1}^{n} \chi_\nu \kappa_{\nu,hk} \qquad (h,k = 1,2,\ldots,m) .$$

Since, by the hypothesis, $\kappa(\alpha) \neq 0$, at least one of the numbers $\kappa_\nu(\alpha)$ is distinct from zero; assume that $\kappa_1(\alpha) \neq 0$.

The $mn$ expressions

$$\phi_{h\nu} = \kappa_\nu f_h'(z) - \sum_{k=1}^{m} \kappa_{\nu,hk} f_k(z) \qquad (h = 1,2,\ldots,m;\ \nu = 1,2,\ldots,n)$$

can by the hypothesis be expanded into power series with coefficients in $K$. By the differential equations $Q$, these series satisfy the identities

$$\sum_{\nu=1}^{n} \chi_\nu \phi_{h\nu} = 0 \qquad (h = 1,2,\ldots,m).$$

Since $\chi_1, \ldots, \chi_n$ are linearly independent over $K$, it follows then that also

$$\phi_{h\nu} = 0 \qquad (h = 1,2,\ldots,m;\ \nu = 1,2,\ldots,n).$$

On choosing here $\nu = 1$, we find that $f(z)$ is a solution of the system

$$Q_1 : w_h' = \sum_{k=1}^{m} q_{1,hk} w_k \qquad (h = 1,2,\ldots,m)$$

where the new coefficients

$$q_{1,hk} = \kappa_1^{-1} \cdot \kappa_{1,hk} \qquad (h,k = 1,2,\ldots,m)$$

are elements of $K(z)$. Since $\kappa_1(\alpha) \neq 0$, none of these coefficients has a pole at $z = \alpha$, and hence $Q_1$ has the required form.

**88.** A further simplification of Lemma (42) concerns the two integers $D(z) = D_{K(z)}\{f_1(z),\ldots,f_m(z)\}$ and $D(\alpha) = D_k\{f_1(\alpha),\ldots,f_m(\alpha)\}$, for which we prove now the formulae

(44): $$D(z) = D_{\mathbb{C}(z)}\{f_1(z),\ldots,f_m(z)\}$$

and

(45): $$D(\alpha) = D_{\mathbb{Q}}\{f_1(\alpha),\ldots,f_m(\alpha)\}.$$

Proof of (44). It evidently suffices to prove that, if $1 \leq M \leq m$, and if $f_1(z), \ldots, f_M(z)$ are algebraically $H$-dependent over $\mathbb{C}(z)$, then they are also algebraically $H$-dependent over $K(z)$; for it is trivial that conversely the $H$-dependence over $K(z)$ implies that over $\mathbb{C}(z)$.

Let then $P(w_1,\ldots,w_M) \not\equiv 0$ be any $H$-polynomial with coefficients in $\mathbb{C}(z)$ such that

(46): $$P\big(f_1(z),\ldots,f_M(z)\big) \equiv 0$$

identically in $z$. This identity may still be multiplied by any polynomial in $\mathbb{C}[z]$ which is not identically zero. Therefore, without loss of generality, the coefficients of $P$ are polynomials in $\mathbb{C}[z]$. Now the total number of coefficients of these polynomials is finite. It follows that there exist

(i) finitely many numbers $\zeta_1, \ldots, \zeta_p$ in $\mathbb{C}$ which are linearly independent over $K$, and

(ii) an equal number of $H$-polynomials

$$P_\pi(w_1,\ldots,w_M) \not\equiv 0 \qquad (\pi = 1,2,\ldots,p)$$

with coefficients in $K[z]$, such that

$$P(w_1,\ldots,w_M) = \sum_{\pi=1}^{p} \zeta_\pi P_\pi(w_1,\ldots,w_M) .$$

The expressions

$$P_\pi\big(f_1(z),\ldots,f_M(z)\big) \text{ , } = \psi_\pi \text{ say,}$$

can be developed into power series in $z$ with coefficients in $K$, and they satisfy the identity

$$\sum_{\pi=1}^{p} \zeta_\pi \psi_\pi = 0 .$$

Since $\zeta_1, \ldots, \zeta_p$ are linearly independent over $K$, this requires the $p$ independent identities

$$P_\pi\big(f_1(z),\ldots,f_M(z)\big) \equiv 0 \qquad (\pi = 1,2,\ldots,p) ,$$

whence the assertion.

Proof of (45). Let $1 \le M \le m$. It suffices to show that if $f_1(\alpha), \ldots, f_M(\alpha)$ are algebraically $H$-dependent over $K$, then they are also algebraically $H$-dependent over $\mathbb{Q}$; for the converse dependence relation is again trivial.

Let $K = \mathbb{Q}(\theta)$, and let $\theta = \theta^{(0)}, \theta^{(1)}, \ldots, \theta^{(N-1)}$ be the algebraic conjugates of $\theta$. We are assuming that there exists an $H$-polynomial $P(w_1,\ldots,w_M) \not\equiv 0$ with coefficients in $K$ such that

$$P\big(f_1(z),\ldots,f_M(z)\big) \equiv 0$$

identically in $z$. This $H$-polynomial can be written in the form

$$P(w_1,\dots,w_M) = \Pi(\theta,w_1,\dots,w_M)$$

where $\Pi(w,w_1,\dots,w_M)$ is a polynomial in $\mathrm{Q}[w,w_1,\dots,w_M]$ which is homogeneous in $w_1, \dots, w_M$. Put

$$P^*(w_1,\dots,w_M) = \prod_{j=0}^{N-1} \Pi\left(\theta^{(j)},w_1,\dots,w_M\right) ,$$

so that $P^*$, by its symmetry in the conjugates of $\theta$, is an $H$-polynomial with coefficients in $\mathrm{Q}$. Evidently also $P^* \not\equiv 0$, and further $P^*\left(f_1(\alpha),\dots,f_M(\alpha)\right) = 0$, whence the assertion.

89. Let us finally combine Lemma (42) with the properties just proved in §§86-88. We thus arrive at the following FIRST MAIN RESULT OF SHIDLOVSKI.

**THEOREM 9.** *Let* $m \geq 2$ ; *let*

$$Q : w_h' = \sum_{k=1}^{m} q_{hk}w_k \qquad (h = 1,2,\dots,m)$$

*be a system of homogeneous linear differential equations with coefficients* $q_{hk}$ *in* $\mathrm{C}(z)$ ; *and let* $\mathrm{f}(z)$ *be a solution of* $Q$ *which has components* $f_1(z), \dots, f_m(z)$ *that are E-functions. Further, let* $\alpha \neq 0$ *be any algebraic number which is not a pole of any* $q_{hk}$ .

*Then the largest number of functions* $f_1(z), \dots, f_m(z)$ *that are algebraically H-independent over* $\mathrm{C}(z)$ *is equal to the largest number of function values* $f_1(\alpha), \dots, f_m(\alpha)$ *that are algebraically H-independent over* $\mathrm{Q}$ .

It is remarkable that there is in this theorem no restriction on the system $Q$ other than that its coefficients $q_{hk}$ have to be rational functions. But the assumption that the components of the solution $\mathrm{f}(z)$ are $E$-functions naturally imposes a very stringent indirect restriction on $Q$ .

Theorem 9 has as an analogue the following SECOND MAIN RESULT OF SHIDLOVSKI which is perhaps even more striking.

**THEOREM 10.** *Let*

$$Q^* : w_h' = q_{h0} + \sum_{k=1}^{m} q_{hk}w_k \qquad (h = 1,2,\dots,m)$$

*be a system of homogeneous or inhomogeneous linear differential equations with coefficients* $q_{hk}$ *in* $\mathrm{C}(z)$ , *and let* $\mathrm{f}(z)$ *be a solution of* $Q^*$ *the* *components* $f_1(z), \ldots, f_m(z)$ *of which are* $E$*-functions. Further, let* $\alpha \neq 0$ *be an algebraic number which is not a pole of any* $q_{hk}$.

*Then the largest number of functions* $f_1(z), \ldots, f_m(z)$ *that are algebraically independent over* $\mathbb{C}(z)$ *is equal to the largest number of function values* $f_1(\alpha), \ldots, f_m(\alpha)$ *that are algebraically independent over* $\mathbb{Q}$.

Proof. We associate with $Q^*$ the following system of homogeneous linear differential equations,

$$Q_H : w_0' = 0 \,, \quad w_h' = q_{h0}w_0 + \sum_{k=1}^{m} q_{hk}w_k \qquad (h = 1,2,\ldots,m) \,.$$

This system deals with $m+1 \geq 2$ functions $w_0, w_1, \ldots, w_m$ ; it has the special solution with the components $f_0(z) \equiv 1$, $f_1(z), \ldots, f_m(z)$, where also $f_0(z)$ is a (trivial) $E$-function. By the final remark in §77, it may be supposed that a maximal $H$-independent subset of $f_0(z), \ldots, f_m(z)$ over $\mathbb{C}(z)$ [or of $f_0(\alpha), \ldots, f_m(\alpha)$ over $\mathbb{Q}$ ] includes $f_0(z)$ [or $f_0(\alpha)$ ].

By formula (4) of §77, the $M+1$ functions $f_0(z), \ldots, f_M(z)$, where $M \leq m$, are exactly then algebraically $H$-independent if $f_1(z), \ldots, f_M(z)$ are algebraically independent, and similarly for the values of these functions at $z = \alpha$. Hence Theorem 10 is contained in Theorem 9.

It may be well to stress once more what is unsatisfactory with both Theorem 9 and Theorem 10. These two theorems apply only to such solutions of linear differential equations which have components that are *E-functions*. On the other hand, there are no *explicit* restrictions on the coefficients $q_{hk}$, apart from the demand that these be rational functions of $z$. Unfortunately, the requirement that the differential equations allow a solution with components that are $E$-functions imposes a very stringent *implicit* restriction on the $q_{hk}$.

The greater part of the proofs of the two Theorems 9 and 10 is independent of the restriction to $E$-functions, and this restriction arose only in the preceding chapter, in connection with Siegel's method based on Diophantine approximations. Thus the problem arises to find a more general theory which applies to solutions of differential equations where the components have power series with algebraic coefficients, but are not necessarily $E$-functions or even entire functions. At present, no such theory is known.

**90.** Shidlovski's two main results imply a number of consequences which have interest in themselves.

Let again

$$Q^* : w_h' = q_{h0} + \sum_{k=1}^{m} q_{hk}w_k \qquad (h = 1,2,\ldots,m)$$

be a (homogeneous or inhomogeneous) system of linear differential equations with rational coefficients, and let $\mathbf{f}(z)$ be a solution with $E$-function components. Let, say, $f_1(z), \ldots, f_M(z)$ be transcendental $E$-functions, while $f_{M+1}(z), \ldots, f_m(z)$ are algebraic $E$-functions; that is, are polynomials. It is immediately clear that the $M$-vector with $f_1(z), \ldots, f_M(z)$ as components satisfies a system of differential equations of the same form as $Q^*$, also with rational coefficients. We may therefore, without loss of generality, always assume that all the components of $\mathbf{f}(z)$ are *transcendental* functions.

If all the components of $\mathbf{f}(z)$ are algebraically independent over $\mathbf{C}(z)$, it follows from Theorem 10 that, for all but finitely many algebraic points $z = \alpha$, the $m$ function values $f_1(\alpha), \ldots, f_m(\alpha)$ are algebraically independent over $\mathbf{Q}$ and so, in particular, are transcendental.

Let this simple case now be excluded. As before, denote by

$$d = d_{\mathbf{C}(z)}\left(f_1(z),\ldots,f_m(z)\right)$$

the maximum number of components of $\mathbf{f}(z)$ that are algebraically independent over $\mathbf{C}(z)$. Then

$$1 \le d \le m-1 .$$

If $\alpha \neq 0$ is algebraic and not a pole of any coefficient $q_{hk}$, then, by Theorem 10, $d$ is also equal to the maximum number of function values $f_1(\alpha), \ldots, f_m(\alpha)$ that are algebraically independent over $\mathbf{Q}$.

Denote by $A$ the set of all algebraic numbers $\alpha \neq 0$ which are not poles of any $q_{hk}$ and for which at least one component of $\mathbf{f}(\alpha)$ is algebraic. Let $\alpha$ be an arbitrary element of $A$, and let the suffix $M = M(\alpha)$ be such that $f_M(\alpha)$ is algebraic. By what has just been said about $d$, it is possible to choose $d$ suffixes $h_1, \ldots, h_d$ all distinct from $M$ and satisfying

$$1 \le h_1 < h_2 < \ldots < h_d \le m$$

such that the function values $f_{h_1}(\alpha), \ldots, f_{h_d}(\alpha)$ are algebraically independent over $\mathbf{Q}$. We assert that this implies that the corresponding functions

$f_{h_1}(z), \ldots, f_{h_d}(z)$ are algebraically independent over $\mathbb{C}(z)$ .

For by the considerations in §88, there would otherwise exist an algebraic number field $K$ and a polynomial $P(w_1,\ldots,w_d) \not\equiv 0$ , with relatively prime coefficients in $K[z]$ , such that

$$P\left(f_{h_1}(z),\ldots,f_{h_d}(z)\right) \equiv 0$$

identically in $z$ . However, on putting here $z = \alpha$ , it would follow that the function values $f_{h_1}(\alpha), \ldots, f_{h_d}(\alpha)$ are algebraically dependent over $K$ and hence also over $\mathbb{Q}$ , contrary to the hypothesis.

91. On the other hand, the $d + 1$ functions $f_{h_1}(z), \ldots, f_{h_d}(z)$ , $f_M(z)$ are algebraically dependent over $\mathbb{C}(z)$ . By §88, they are then also algebraically dependent over $K^*(z)$ where $K^*$ denotes a certain algebraic number field. Hence there exists an irreducible polynomial $R(w_1,\ldots,w_d,w,z) \not\equiv 0$ in the polynomial ring $K^*[w_1,\ldots,w_d,w,z]$ such that

(47): $$R\left(f_{h_1}(z),\ldots,f_{h_d}(z),f_M(z),z\right) \equiv 0$$

identically in $z$ . Since $f_M(z)$ is a transcendental function of $z$ , $R$ depends on at least one of the indeterminates $w_1, \ldots, w_d$ .

In explicit form, $R$ can be written as

(48): $$R(w_1,\ldots,w_d,w,z) = \sum_{(j)} R_{j_1\ldots j_d}(w,z)w_1^{j_1} \ldots w_d^{j_d} .$$

Here the summation extends over finitely many distinct sets of $d$ non-negative integers $(j) = (j_1,\ldots,j_d)$ , and the $R_{j_1\ldots j_d}(w,z)$ are polynomials in $K^*[w,z]$ which do not vanish identically and are relatively prime. From the hypothesis, $R$ is irreducible and not a constant; hence the right-hand side in (48) is a sum of at least two terms.

For $z = \alpha$ , the identity (47) changes into the equation

(49): $$R\left(f_{h_1}(\alpha),\ldots,f_{h_d}(\alpha),f_M(\alpha),\alpha\right) = 0 .$$

Here $\alpha$ and $f_M(\alpha)$ , hence also all the coefficients $R_{j_1\ldots j_d}\left(f_M(\alpha),\alpha\right)$ , assume algebraic values, while by hypothesis $f_{h_1}(\alpha), \ldots, f_{h_d}(\alpha)$ are algebraically independent over $\mathbb{Q}$ . The equation (49) requires therefore that

(50): $$R_{j_1\cdots j_d}\left(f_M(\alpha),\alpha\right) = 0 \quad \text{for all systems} \quad (j) .$$

Since $K^*[w,z]$ is a polynomial ring over a field, the law of unique factorisation holds. There exist then finitely many irreducible polynomials

$$p_1(w,z), \ldots, p_r(w,z) ,$$

where no two of these polynomials differ only by a constant factor, such that each coefficient $R_{j_1\cdots j_d}(w,z)$ can be written as a product

$$R_{j_1\cdots j_d}(w,z) = c_{j_1\cdots j_d} \prod_{\rho=1}^{r} p_\rho(w,z)^{e_{\rho j_1\cdots j_d}} .$$

Here the factors $c_{j_1\cdots j_d}$ are nonzero elements of $K^*$, and the exponents $e_{\rho j_1\cdots j_d}$ are non-negative integers; by (50),

$$\sum_{\rho=1}^{r} e_{\rho j_1\cdots j_d} > 0$$

for all systems $(j)$.

Each of the equations (50) implies at least one equation

$$p_\rho\left(f_M(\alpha),\alpha\right) = 0 .$$

In fact, at least two such equations must hold, say the equations

(51): $$p_1\left(f_M(\alpha),\alpha\right) = 0 , \quad p_2\left(f_M(\alpha),\alpha\right) = 0 ;$$

for (48) contains at least two terms, and the polynomials $R_{j_1\cdots j_d}(w,z)$ in their entirety are relatively prime.

Since $p_1(w,z)$ and $p_2(w,z)$ are irreducible and essentially distinct, their resultant relative to $w$, $\Omega(z)$ say, is a polynomial in $K^*[z]$ which does not vanish identically. By the definition of the resultant, every solution $\alpha$ of (51) satisfies the equation

$$\Omega(\alpha) = 0 .$$

On retracing the proof, it follows that (49) can be satisfied for at most finitely many algebraic values of $\alpha$, and hence that the set $A$ contains at most finitely many elements that belong to the system of suffixes $M, h_1, \ldots, h_d$. Since there are only finitely many such systems, it follows that $A$ is a finite set.

The result so obtained may be formulated as follows.

(52): *Let* f$(z)$ *have components which are E-functions and let it be a solution* *of a system $Q^*$ of differential equations. Each component of* $f(z)$ *which is a transcendental function assumes an algebraic value at not more than finitely algebraic points* $z = \alpha$ .

This result implies the following assertion made in the introduction (§76).

(53): *Every transcendental Siegel E-function is algebraic at not more than finitely many algebraic points* $z = \alpha$ .

92. The following application of Theorem 10 and of (52) has some interest. Let $p_0(z) \not\equiv 0$ , $p_1(z), \ldots, p_m(z)$ , $p(z)$ be $m+2$ polynomials in $C[z]$ , and let $f(z)$ be an $E$-function satisfying the linear differential equation

$$L^* : p_0(z)w^{(m)} + p_1(z)w^{(m-1)} + \ldots + p_m(z)w + p(z) = 0 .$$

On putting

$$w_1 = w,\ w_2 = w',\ \ldots,\ w_m = w^{(m-1)} ,$$

$L^*$ can be written as the system of first order differential equations,

$$Q^* : w_1' = w_2,\ w_2' = w_3,\ \ldots,\ w_m' = -p_0^{-1}(p_1w_m+p_2w_{m-1}+\ldots+p_mw_1+p) .$$

In addition to $L^*$ , $f(z)$ may satisfy some algebraic differential equation of lower order than $m$ . Let $M$ be the smallest integer satisfying

$$1 \le M \le m-1$$

for which there exists a polynomial $P(w_1,w_2,\ldots,w_{M+1},z) \not\equiv 0$ with coefficients in $C$ such that

(54): $$P\big(f(z),f'(z),\ldots,f^{(M)}(z),z\big) \equiv 0$$

identically in $z$ . Then $P$ depends on the indeterminate $w_{M+1}$ , and upon differentiating (54) repeatedly with respect to $z$ , we obtain new relations which determine $f^{(M+1)}(z), \ldots, f^{(m-1)}(z)$ rationally in terms of $f(z), f'(z), \ldots, f^{(M)}(z)$ and $z$ . It follows that exactly $M$ and no more of the functions

$$f(z),\ f'(z),\ \ldots,\ f^{(m-1)}(z)$$

are algebraically independent over $C(z)$ . This remains true, with $M = m$ , if $f(z)$ does not satisfy any algebraic differential equation of order less than $m$ .

Theorem 10 leads therefore immediately to the following lemma.

(55): *Let* $f(z)$ *be an E-function satisfying the linear differential equation* $L^*$ . *Let* $M$ *be the smallest positive integer such that* $f(z)$ *satisfies* *an algebraic differential equation* (54) *of order* $M$ *, and let* $\alpha \neq 0$ *be an algebraic number for which* $p_0(\alpha) \neq 0$ .

*Then certain* $M$ *among the function values*

$$f(\alpha),\ f'(\alpha),\ \ldots,\ f^{(m-1)}(\alpha)$$

*and no more are algebraically independent over* $Q$ .

**93.** The theorems and lemmas proved up to now in this chapter allow many applications. Such applications to special functions will be considered in the next chapter; but it is perhaps useful to mention already a few more general remarks.

Again let $f(z)$ be any transcendental Siegel $E$-function. Then, by (53), $f(\alpha)$ is transcendental for all but finitely many algebraic numbers $\alpha$ . It follows that *if* $\beta$ *is any algebraic number, then the equation*

$$f(\alpha) = \beta ,$$

which in general has infinitely many solutions, *has at most finitely many algebraic solutions* $\alpha$ . In particular, there can be at most *finitely many algebraic zeros of* $f(z)$ .

Next consider finitely many Siegel $E$-functions $f_1(z),\ \ldots,\ f_M(z)$ and a polynomial $P(w_1,\ldots,w_M,z)$ with algebraic coefficients. The expression

$$P\big(f_1(z),\ldots,f_M(z),z\big)\ , \quad = f(z) \quad \text{say},$$

likewise is a Siegel $E$-function. Assume that $f(z)$ does not reduce to a polynomial in $z$ , so that it is a transcendental function. Lemma (53) then shows that $f(\alpha)$ is algebraic for at most finitely many algebraic values of $\alpha$ . In particular, $f(z)$ *cannot have infinitely many algebraic zeros.*

**94.** Consider again a solution $\mathbf{f}(z)$ of the differential equations

$$Q^* : w_h' = q_{h0} + \sum_{k=1}^{m} q_{hk} w_k \qquad (h = 1,2,\ldots,m)$$

with components $f_1(z),\ \ldots,\ f_m(z)$ that are $E$-functions, say relative to the algebraic number field $K$ . As before, denote by

$$d = d_{K(z)}\big(f_1(z),\ldots,f_m(z)\big) = d_{\mathrm{C}(z)}\big(f_1(z),\ldots,f_m(z)\big)$$

the maximum number of components of $\mathbf{f}(z)$ which are algebraically independent over $K(z)$ , and thus, by a proof similar to that of (44), are also algebraically independent over $\mathrm{C}(z)$ . The notation may be chosen so that $f_1(z),\ \ldots,\ f_d(z)$ are algebraically independent over $K(z)$ , but that $f_1(z),\ \ldots,\ f_d(z),\ f_k(z)$ are algebraically dependent over $K(z)$ if $d+1 \le k \le m$ .

Put $L = K(z)$ and denote by $U$ and $V$ the two extension fields

$$U = L\{f_1(z),\dots,f_d(z)\} \quad \text{and} \quad V = L\{f_1(z),\dots,f_m(z)\} = U\{f_{d+1}(z),\dots,f_m(z)\} .$$

Just as in §78, there exists an element of $V$ , say $g(z)$ , such that $V$ is the extension field

$$V = U\{g(z)\}$$

of $U$ . Here $g(z)$ is algebraic over $U$ , and without loss of generality, its minimal equation has the form

(56): $$A\{g(z);f_1(z),\dots,f_d(z),z\} = 0 ,$$

where

(57): $$A(w;w_1,\dots,w_d,z) = w^c + \sum_{h=1}^{c} C_h(w_1,\dots,w_d,z)w^{c-h}$$

is an irreducible polynomial in $K[w,w_1,\dots,w_d,z]$ which is monic and of positive degree $c$ in $w$ .

The $m - d$ functions $f_{d+1}(z)$, ..., $f_m(z)$ all lie in $V$ . They can therefore be written in the form

(58): $$f_k(z) = \frac{\sum_{h=0}^{c-1} D_{hk}\{f_1(z),\dots,f_d(z),z\}g(z)^h}{D_0\{f_1(z),\dots,f_d(z),z\}} \qquad (k = d+1,d+2,\dots,m) .$$

Here the expressions $D_{hk}(w_1,\dots,w_d,z)$ and $D_0(w_1,\dots,w_d,z)$ are polynomials in $K[w_1,\dots,w_d,z]$ , and the denominator polynomial $D_0(w_1,\dots,w_d,z)$ is not identically zero. Hence also

(59): $$D_0\{f_1(z),\dots,f_d(z),z\} \not\equiv 0 .$$

Put

$$f(z) = D_0\{f_1(z),\dots,f_d(z),z\} ;$$

then $f(z)$ is a Siegel $E$-function which does not vanish identically.

Finally denote again by $\alpha$ any algebraic number which is not a pole of any one of the rational functions $q_{hk}$ . By Theorem 10, exactly $d$ of the function values $f_1(\alpha)$, ..., $f_m(\alpha)$ are algebraically independent over $Q$ and hence also over $K$ . Assume, however, that the first $d$ function values $f_1(\alpha)$, ..., $f_d(\alpha)$ are not all algebraically independent, hence that

$$\delta = d_K\{f_1(\alpha),\ldots,f_d(\alpha)\} < d .$$

Here the case when $\delta = 0$ is not excluded. Without loss of generality, we may further assume that $f_1(\alpha), \ldots, f_\delta(\alpha)$ are algebraically independent over $K$ if $\delta > 0$, but that all $d$ function values $f_1(\alpha), \ldots, f_d(\alpha)$ are algebraic if $\delta = 0$.

Denote by $K^*$ the transcendental extension field $K\{f_1(\alpha),\ldots,f_\delta(\alpha)\}$ if $\delta > 0$ and the algebraic number field $K$ if $\delta = 0$. On putting $z = \alpha$, the equations (56) and (57) show that $g(\alpha)$ lies in an algebraic extension field of $K\{f_1(\alpha),\ldots,f_d(\alpha)\}$; therefore, by the definition of $\delta$, $g(\alpha)$ is also algebraic over $K^*$. Similarly, all the values $D_{hk}\{f_1(\alpha),\ldots,f_d(\alpha),\alpha\}$ and $D_0\{f_1(\alpha),\ldots,f_d(\alpha),\alpha\}$ are algebraic over $K^*$.

It follows therefore from (58) that one of the following two cases must always hold:

Case A: $f(\alpha) = 0$, or

Case B: *the* $m - d$ *function values* $f_{d+1}(\alpha), \ldots, f_m(\alpha)$ *are algebraic over* $K^*$.

Here Case B can immediately be excluded, because it would imply that *fewer than* $d$ of the function values $f_1(\alpha), \ldots, f_m(\alpha)$ are algebraically independent over $K$, contrary to the definition of $d$. Hence Case A must hold. Here $f(z) \not\equiv 0$ is a Siegel $E$-function. If it degenerates into a polynomial, it has at most finitely many zeros; and if it is a transcendental function, it still has by §93 at most finitely many *algebraic* zeros.

We have therefore proved the following result.

**(60)**: *Let* $\mathbf{f}(z)$ *be a solution of*

$$Q^* : w_h' = q_{h0} + \sum_{k=1}^{m} q_{hk}w_k \qquad (h = 1,2,\ldots,m),$$

*where the* $q_{hk}$ *are in* $\mathbb{C}(z)$, *and where the components of* $\mathbf{f}(z)$ *are* $E$*-functions. Let exactly* $d$ *of these components,* $f_1(z), \ldots, f_d(z)$ *say, be algebraically independent over* $\mathbb{C}(z)$. *Then, for all but finitely many algebraic numbers* $\alpha$, *the values* $f_1(\alpha), \ldots, f_d(\alpha)$ *are algebraically independent over* $\mathbb{Q}$.

An immediate consequence of this lemma is as follows.

**(61)**: *Let* $f_1(z), \ldots, f_M(z)$ *be finitely many Siegel* $E$*-functions which are* *algebraically independent over* $\mathbb{C}(z)$ . *Then there exist at most finitely many algebraic numbers* $\alpha$ *for which the values* $f_1(\alpha), \ldots, f_M(\alpha)$ *are algebraically dependent over* $\mathbb{Q}$ .

For each function is the component of a vector solution of a suitable system $Q^*$ of linear differential equations. By combining all these separate systems into one system, while at the same time combining the vector solutions into one big vector, we come back to the result (60), except that $f_1(z), \ldots, f_M(z)$ will now be only a subsystem of $f_1(z), \ldots, f_d(z)$ . But this change will naturally not affect the result.

# CHAPTER 7

## APPLICATIONS OF SHIDLOVSKI'S MAIN THEOREMS TO SPECIAL FUNCTIONS

**95.** Shidlovski's Theorems 9 and 10 allow many interesting applications to problems of transcendency. However, these theorems apply only to Siegel $E$-functions, thus to a very restricted class of entire functions.

This restriction is, in fact, even more severe because one still does not know the most general Siegel $E$-function. We shall therefore consider here only a special class of hypergeometric functions which were proved by Siegel (1929 and 1949) both to be $E$-functions, and to satisfy linear differential equations with rational functions as coefficients. All Siegel $E$-functions so far known can be derived from these hypergeometric $E$-functions by a change of variable, or by algebraic field operations, or by integrations over $z$, or by partial differentiations with respect to the parameters.

Both the exponential function $e^z$ and the Bessel function $K_0(z;\nu)$ of Siegel are examples of such hypergeometric $E$-functions. Shidlovski's theorems may then be applied to these functions, yielding the classical results by Hermite and Lindemann, and by Siegel, respectively.

**96.** We begin by proving an arithmetic lemma which goes back to A. Thue and W. Maier. It allows one to study the arithmetic character of the hypergeometric $E$-functions.

Let $\alpha = a/b$ be a rational number distinct from $0, -1, -2, \ldots$, where $a$ and $b$ are integers satisfying $(a,b) = 1$ and $b > 0$. Write $c = |a| + b$, so that $c \geq 1$, and define expressions $[\alpha,n]$ by the formulae

$$[\alpha,0] = 1\ , \quad [\alpha,n] = \alpha(\alpha+1)\ \ldots\ (\alpha+n-1) = a(a+b)\ \ldots\ \big(a+(n-1)b\big)b^{-n} \quad \text{for } n \geq 1\ ;$$

hence, in particular, $[1,n] = n!$.

Assume that $n \geq 1$. In the prime factorisations

$$b = \prod_p p^{j_p}\ , \quad [\alpha,n] = \pm\prod_p p^{m_p}\ , \quad n! = \prod_p p^{n_p}\ ,$$

where $p$ runs over all distinct primes, $j_p$ and $n_p$ are non-negative, while $m_p$ may be positive, negative, or zero.

Suppose, firstly, that $p$ is a prime which does not divide $b$, so that

$m_p \geq 0$ . For every positive integer $t$ , exactly one of any $p^t$ consecutive elements of the sequence

$$a,\ a+b,\ a+2b,\ \ldots$$

is divisible by $p^t$ . Of the factors

$$a,\ a+b,\ \ldots,\ a+(n-1)b$$

in the numerator of $[\alpha,n]$ , therefore, at least $[n/p^t]$ and at most $[n/p^t]+1$ are divisible by $p^t$ ; here $[x]$ is as usual the integral part of $x$ . Since $p^t$ divides one of the factors $a, a+b, \ldots, a+(n-1)b$ , we must have

$$\begin{aligned} p^t &\leq \max_{0\leq l\leq n-1} |a+lb| = \max(|a|,|a+(n-1)b|) \\ &= |a| + (n-1)b < n|a| + nb = nc \ . \end{aligned}$$

Hence $p^t \leq cn$ and thus $t \leq M_p$ , where $M_p = [\log(cn)/\log p]$ . It follows that

$$\sum_{t=1}^{M_p} [n/p^t] \leq m_p \leq \sum_{t=1}^{M_p} ([n/p^t]+1) \quad \text{if} \quad p \nmid b \ .$$

If, secondly, $p$ is a prime factor of $b$ , this formula no longer holds. On account of the denominator $b^n$ of $[\alpha,n]$ now

$$m_p = -j_p n \quad \text{if} \quad p \mid b \ .$$

In the special case of $[1,n] = n!$ the same proof leads to the stronger result that for all $p$ ,

$$n_p = \sum_{t=1}^{M_p} [n/p^t] \ ,$$

hence that

$$0 \leq n_p \leq \sum_{t=1}^{\infty} n.2^{-t} = n \ .$$

Therefore

$$0 \leq m_p - n_p \leq M_p \quad \text{if} \quad p \nmid b \ ,$$

$$0 \geq m_p - n_p \geq -(j_p+1)n \geq -2j_p n \quad \text{if} \quad p \mid b \ .$$

Now put

$$U_n = \prod_{p\nmid b} p^{M_p} \ , \quad V_n = b^{2n} \ ,$$

and write $[\alpha,n]/n!$ as a reduced fraction

$$[\alpha,n]/n! = u_n/v_n \qquad (u_0 = v_0 = 1) ,$$

where the integers $u_n$ and $v_n > 0$ are relatively prime. It follows then that

$$u_n \mid U_n \text{ and } v_n \mid V_n .$$

Further, it is obvious from the definitions that

$$U_n \mid U_{n+1} \text{ and } V_n \mid V_{n+1} .$$

Hence the least common multiple of $u_0, u_1, \ldots, u_n$ is a divisor of $U_n$, and that of $v_0, v_1, \ldots, v_n$ is a divisor of $V_n$.

Finally, by a weak form of the prime number theorem,

$$\log U_n \leq \sum_{p \leq cn} \frac{\log(cn)}{\log p} \log p = O\left(\frac{cn}{\log(cn)}\right) \log(cn) = O(n) .$$

Hence there exists a constant integer $C$ which depends only on $\alpha$ and which without loss of generality is greater than or equal to $b^2$ such that

$$U_n \leq C^n \text{ and } V_n \leq C^n .$$

The result so proved may be formulated as a lemma.

**(1):** *Let* $\alpha = a/b$, *where* $(a,b) = 1$ *and* $b \geq 1$, *be a rational number distinct from* $0, -1, -2, \ldots$. *There exists a positive integer* $C$ *with the following property.*

*For any integer* $n \geq 0$ *write* $[\alpha,n]/n!$ *as a reduced fraction* $u_n/v_n$ *where* $v_n > 0$ *and* $(u_n,v_n) = 1$. *Then the least common multiples of* $u_0, u_1, \ldots, u_n$ *and of* $v_0, v_1, \ldots, v_n$ *are not larger than* $C^n$. *In particular,* $|u_n|$ *and* $v_n$ *do not exceed* $C^n$.

97. The lemma just proved can be generalised as follows.

**(2):** *Let* $\alpha_1, \ldots, \alpha_s, \beta_1, \ldots, \beta_s$ *be* $2s$ *rational numbers distinct from* $0, -1, -2, \ldots$. *There exists a positive integer* $C_1$ *with the following property. For* $n = 0, 1, 2, \ldots$, *let*

$$\frac{[\alpha_1,n]\ldots[\alpha_s,n]}{[\beta_1,n]\ldots[\beta_s,n]} = x_n/y_n \qquad (x_0 = y_0 = 1)$$

*where* $x_n \neq 0$ *and* $y_n \neq 0$ *are integers that are relatively prime. Then the least common multiples of* $x_0, x_1, \ldots, x_n$ *and of* $y_0, y_1, \ldots, y_n$ *do not exceed* $C_1^n$ *, and in particular* $|x_n| \le C_1^n$ *and* $|y_n| \le C_1^n$ *.*

For

$$x_n/y_n = \prod_{h=1}^{s} \left( \frac{[\alpha_h,n]}{n!} \cdot \frac{n!}{[\beta_h,n]} \right) ,$$

so that the assertion follows from (1).

We also need the following well known lemma.

**(3):** *If* $n$ *and* $t$ *are positive integers, the quotient* $(nt)!/\{n!\}^t$ *is an integer not greater than* $t^{nt}$ *.*

For $(nt)!/\{n!\}^t$ is the coefficient of $a_1^n a_2^n \ldots a_t^n$ in the multinomial expansion

$$(a_1+a_2+\ldots+a_t)^{nt} = \sum \frac{(nt)!}{h_1!h_2!\ldots h_t!} a_1^{h_1} a_2^{h_2} \ldots a_t^{h_t}$$

where the summation extends over all sets of $t$ integers $h_1, h_2, \ldots, h_t$ satisfying $h_1 \ge 0, h_2 \ge 0, \ldots, h_t \ge 0, h_1 + h_2 + \ldots + h_t = nt$ . This coefficient is therefore an integer. On putting $a_1 = a_2 = \ldots = a_t = 1$ , it follows that it satisfies the asserted inequality.

**98.** Let $r$ and $s$ be two integers satisfying $0 \le r < s$ , and let $t = s - r$ , so that $t \ge 1$ . Denote by $\alpha_1, \ldots, \alpha_r, \beta_1, \ldots, \beta_s$ any $r + s$ rational numbers distinct from $0, -1, -2, -3, \ldots$ . The hypergeometric functions to be considered are of the form

$$f(z) = \sum_{n=0}^{\infty} \frac{[\alpha_1,n]\ldots[\alpha_r,n]}{[\beta_1,n]\ldots[\beta_s,n]} (z/t)^{nt} ;$$

here the numerators of the coefficients are to mean 1 if $r = 0$ . The series is formed in terms of $z/t$ rather than of $z$ because this will simplify the differential equation for $f(z)$ which will be established in the next section. The hypothesis implies that the numerators and the denominators of all the coefficients of $f(z)$ are distinct from zero; hence $f(z)$ does not reduce to a polynomial.

Put $\alpha_{r+1} = \ldots = \alpha_s = 1$ . Then $f(z)$ can also be written as

$$f(z) = \sum_{\nu=0}^{\infty} f_\nu \frac{z^\nu}{\nu!} ,$$

where we have put

$$f_\nu = \begin{cases} \dfrac{[\alpha_1,n]\ldots[\alpha_s,n]}{[\beta_1,n]\ldots[\beta_s,n]} \dfrac{(nt)!}{(n!)^t t^{nt}} & \text{if } \nu = nt\ , \quad n = 0, 1, 2, \ldots\ , \\ 0 & \text{otherwise.} \end{cases}$$

Now let $\varepsilon$ be an arbitrarily small positive constant. By (2) and (3), the numerator and the denominator of $f_\nu$, for all $\nu = nt$, are integers the absolute values of which are at most

$$C_1^n . t^{nt} = O(\nu^{\varepsilon\nu}) \ .$$

The same estimate holds also for $|f_\nu|$, and for the least common denominator of the $\nu + 1$ coefficients $f_0, f_1, \ldots, f_\nu$. This proves that *$f(z)$ is an E-function relative to* Q. Since $f(z)$ is not a polynomial, it is then necessarily an *entire transcendental function.*

99. We show next that $f(z)$, as just defined, satisfies a linear differential equation with coefficients that are rational functions of $z$; this means that *$f(z)$ is a Siegel E-function.*

Denote by $\theta$ the differential operator

$$\theta = \frac{z}{t}\frac{d}{dz} \ .$$

If $\gamma$ is any constant,

$$(\theta+\gamma)\{(z/t)^{nt}\} = (n+\gamma)(z/t)^{nt} \ .$$

It follows that

$$(\theta+\alpha_1) \ldots (\theta+\alpha_r)\{f(z)\} = \sum_{n=0}^{\infty} (n+\alpha_1) \ldots (n+\alpha_r) \frac{[\alpha_1,n]\ldots[\alpha_r,n]}{[\beta_1,n]\ldots[\beta_s,n]} (z/t)^{nt} =$$

$$= \sum_{n=0}^{\infty} \frac{[\alpha_1,n+1]\ldots[\alpha_r,n+1]}{[\beta_1,n]\ldots[\beta_s,n]} (z/t)^{nt} \ ,$$

and that further

$$(\theta+\beta_1-1) \ldots (\theta+\beta_s-1)\{f(z)\} = \sum_{n=0}^{\infty} (n+\beta_1-1) \ldots (n+\beta_s-1) \frac{[\alpha_1,n]\ldots[\alpha_r,n]}{[\beta_1,n]\ldots[\beta_s,n]} (z/t)^{nt} =$$

$$= (\beta_1-1) \ldots (\beta_s-1) + \sum_{n=0}^{\infty} \frac{[\alpha_1,n+1]\ldots[\alpha_r,n+1]}{[\beta_1,n]\ldots[\beta_s,n]} (z/t)^{nt+t} \ .$$

*Therefore $w = f(z)$ satisfies the linear differential equation*

(4): $$\left((\theta+\beta_1-1)\ldots(\theta+\beta_s-1)-(z/t)^t(\theta+\alpha_1)\ldots(\theta+\alpha_r)\right)w = (\beta_1-1) \ldots (\beta_s-1) \ .$$

This differential equation has the order $s$ ; its coefficients are rational functions, and the equation is homogeneous if and only if at least one of the numbers $\beta_1, \ldots, \beta_s$ is equal to 1 .

It is clear that if the parameters $\alpha_1, \ldots, \alpha_r, \beta_1, \ldots, \beta_s$ are no longer restricted to be rational numbers but are allowed to be arbitrary complex numbers distinct from 0, -1, -2, -3, ... , then $f(z)$ still is an entire transcendental function and still satisfies the differential equation (4); but we naturally cannot now assert that it is an $E$-function.

**100.** From the hypergeometric $E$-functions defined in §98 one can derive other classes of $E$-functions by applying suitable operations. It will suffice to mention the following allowed operations.

(a): $f(z)$ is replaced by $f(\alpha z)$ where $\alpha \neq 0$ is any algebraic number.

(b): $f(z)$ is replaced by $f'(z)$ or by $\int_0^z f(z)dz$ , or more generally by

$$z^{-a} \int_0^z z^{a-1} f(z)dz$$

where $a$ is a nonzero rational number, and where the series is integrated formally term by term.

(c): $f(z)$ is differentiated partially once or more often relative to some of its parameters $\alpha_1, \ldots, \alpha_r, \beta_1, \ldots, \beta_s$ .

All the $E$-functions, $g(z)$ say, which are so obtained satisfy suitable linear differential equations with coefficients in $\mathbf{C}(z)$ . If such an equation is of order $r$ , it can be replaced by a system of $r$ linear differential equations of the first order for the $r$ functions $g(z), g'(z), \ldots, g^{(r-1)}(z)$ . We need thus deal only with systems of functions that satisfy systems of linear differential equations of the *first* order of the types $Q$ and $Q^*$ as studied in the previous chapters.

Theorems 9 and 10 show that in order to decide the arithmetic behaviour of the values of such functions at algebraic points, it suffices to find out how many of these functions are algebraically independent over $\mathbf{C}(z)$ . It is thus necessary to solve the following problem.

*Assume that the E-functions* $f_1(z), \ldots, f_m(z)$ *form a solution of the system of linear differential equations*

$$Q^* : w_h' = q_{h0} + \sum_{k=1}^{m} q_{hk} w_k \qquad (h = 1,2,\ldots,m)$$

*with coefficients in* $C(z)$ . *To determine the largest number of these functions that are algebraically independent over* $C(z)$ .

At present, this problem still presents difficulties, and each case has to be studied separately. Shidlovski (1966) used ideas applied already by Siegel (1929, 1949) and obtained many special results. A more general method was recently introduced by Oleinikov (1969). The next sections will deal with some examples.

**101.** In the simplest case when $r = 0$ , $s = 1$ , $\beta_1 = 1$ , the hypergeometric function $f(z)$ becomes identical with the exponential function $e^z$ . We consider instead the system of $m$ Siegel $E$-functions

$$w_1 = e^{\gamma_1 z}, \ldots, w_m = e^{\gamma_m z}$$

where $\gamma_1 \neq 0, \ldots, \gamma_m \neq 0$ are (real or complex) algebraic numbers. These $m$ functions form a solution of the system of homogeneous linear differential equations

$$Q_\gamma : w_h' = \gamma_h w_h \qquad (h = 1,2,\ldots,m) .$$

We shall now determine conditions for $\gamma_1, \ldots, \gamma_m$ under which these $m$ exponential functions are algebraically independent over $C(z)$ , and for this purpose we begin with the simpler case of linear independence.

(5): *If the algebraic numbers* $\gamma_1, \ldots, \gamma_m$ *are all distinct, then* $e^{\gamma_1 z}, \ldots, e^{\gamma_m z}$ *are linearly independent over* $C(z)$ .

Proof. Let the assertion be false. Then there exists an integer $M$ satisfying $1 \leq M \leq m-1$ such that certain $M$ , but no more, of the functions $e^{\gamma_1 z}, \ldots, e^{\gamma_m z}$ are linearly independent. The notation can then be chosen such that $e^{\gamma_1 z}, \ldots, e^{\gamma_M z}$ , but not $e^{\gamma_1 z}, \ldots, e^{\gamma_M z}, e^{\gamma_m z}$ , are linearly independent over $C(z)$ . Hence there exist $M + 1$ rational functions $r_1(z), \ldots, r_M(z), r_m(z)$ not all identically zero such that

$$(6): \qquad r_1(z)e^{\gamma_1 z} + \ldots + r_M(z)e^{\gamma_M z} + r_m(z)e^{\gamma_m z} \equiv 0 .$$

Here $r_m(z) \not\equiv 0$ , because the first $M$ exponential functions are linearly independent. Further, since $e^{\gamma_m z} \not\equiv 0$ , at least one of the rational functions $r_1(z), \ldots, r_M(z)$ does not vanish identically, say $r_1(z) \not\equiv 0$ .

Upon dividing by $r_m(z)e^{\gamma_m z}$ and applying the differential operator

$$e^{\gamma_m z}\frac{d}{dz}$$

to (6), a new identity of the form

$$r_1^*(z)e^{\gamma_1 z} + \dots + r_M^*(z)e^{\gamma_M z} \equiv 0$$

is obtained, where $r_1^*(z), \dots, r_M^*(z)$ are again rational functions. In particular,

$$r_1^*(z) = (\gamma_1-\gamma_m)\frac{r_1(z)}{r_m(z)} + \frac{d}{dz}\left(\frac{r_1(z)}{r_m(z)}\right) .$$

Since $r_1(z)/r_m(z) \not\equiv 0$, and since by $\gamma_1 \neq \gamma_m$ no solution $w \not\equiv 0$ of

$$w' = (\gamma_m-\gamma_1)w$$

is a rational function, it follows that $r_1^*(z) \not\equiv 0$, hence that $e^{\gamma_1 z}, \dots, e^{\gamma_M z}$ are linearly dependent over $\mathbb{C}(z)$, contrary to the definition of $M$.

(7): *Let $\gamma_1, \dots, \gamma_n$ be finitely many algebraic numbers which are linearly independent over* $\mathbb{Q}$. *Then the exponential functions $e^{\gamma_1 z}, \dots, e^{\gamma_n z}$ are algebraically independent over* $\mathbb{C}(z)$.

Proof. Denote by

$$P(z,w_1,\dots,w_n) = \sum_{(h)\in H} P_{(h)}(z)w_1^{h_1}\dots w_n^{h_n} \not\equiv 0$$

an arbitrary polynomial in $\mathbb{C}[z,w_1,\dots,w_n]$; here $H$ stands for a finite set of distinct systems $(h) = (h_1,\dots,h_n)$ of non-negative integers, and the coefficients $P_{(h)}(z)$ are polynomials not identically zero in $\mathbb{C}[z]$. By the hypothesis, no two of the sums

$$\gamma(h) = h_1\gamma_1 + \dots + h_n\gamma_n$$

belonging to the different elements $(h)$ of $H$ have the same value. The corresponding exponential functions

$$e^{\gamma(h)z}, \text{ where } (h) \in H,$$

are therefore, by (5), linearly independent over $\mathbb{C}(z)$. Since evidently

$$P\left(z,e^{\gamma_1 z},\dots,e^{\gamma_n z}\right) = \sum_{(h)\in H} P_{(h)}(z)e^{\gamma(h)z},$$

it follows then that

$$P\left(z, e^{\gamma_1 z}, \ldots, e^{\gamma_n z}\right) \not\equiv 0 ,$$

whence the assertion.

The $n$ $E$-functions

$$w_1 = e^{\gamma_1 z}, \ldots, w_n = e^{\gamma_n z}$$

form a solution of the system of linear differential equations

$$Q_\gamma : w_h' = \gamma_h w_h \qquad (h = 1,2,\ldots,n)$$

with constant coefficients. Theorem 10 may therefore be applied to these functions, with $\kappa \equiv 1$, and (7) implies the following classical result.

**THEOREM OF HERMITE-LINDEMANN.** *If* $\gamma_1, \ldots, \gamma_n$ *are finitely many real or complex algebraic numbers linearly independent over* $Q$ *, then the function values*

$$e^{\gamma_1}, \ldots, e^{\gamma_n}$$

*are algebraically independent over* $Q$ *. In particular, each one of these values is transcendental.*

When $n = 1$, this theorem implies the transcendency of $e^\gamma$ for algebraic $\gamma \neq 0$, hence the transcendency of $e$ if $\gamma = 1$. It next implies the transcendency of $\pi i$ and thus of $\pi$ because $e^{\pi i} = -1$. It further implies the transcencency of $\log \delta$ for algebraic $\delta$ distinct from $0$ and $1$, and that of other related function values.

**102.** For $r = 0$, $s = 1$, and for arbitrary values of $\beta \neq 0, -1, -2, \ldots$ we obtain the hypergeometric function

$$f(z;\beta) = \sum_{n=0}^{\infty} \frac{1}{\beta(\beta+1)\ldots(\beta+n-1)} z^n$$

which, by (4), satisfies the linear differential equation

$$w' = \frac{\beta-1}{z} + \left(1 - \frac{\beta-1}{z}\right)w .$$

This differential equation has the general integral

$$w = z^{-(\beta-1)}e^z\left((\beta-1)\int z^{\beta-2}e^{-z}dz+c\right) ,$$

where $c$ is an arbitrary constant. Therefore, for $\beta > 0$, $f(z;\beta)$ allows the explicit integral representation

$$f(z;\beta) = 1 + z^{-(\beta-1)}e^{z}\int_0^z z^{\beta-1}e^{-z}dz .$$

For $\beta < 0$ , one can obtain similar representations by making use of the obvious functional equation

$$f(z;\beta) = 1 + \beta^{-1}f(z;\beta+1)z .$$

In particular, $f(z;1) = e^{z}$ , and $f(z;\beta) = \left\{e^{z} - \left(1 + z + \ldots + \frac{z^{\beta-2}}{(\beta-2)!}\right)\right\}\frac{(\beta-1)!}{z^{\beta-1}}$ if $\beta$ is a positive integer.

Shidlovski (1959) proved the following general independence result.

(8): *Let* $\beta_0 \geq 1$ *be an integer, and let* $\beta_1, \ldots, \beta_m$ *be finitely many non-integral rational numbers, no two of which differ by an integer. Let further* $\omega_1, \ldots, \omega_n$ *be finitely many algebraic numbers that are linearly independent over* $Q$ *, and let* $\Omega_1, \ldots, \Omega_n$ *be distinct algebraic numbers not zero. Then the* $(m+1)n$ *functions*

(9): $$f(\omega_k z;\beta_0),\ f(\Omega_k z;\beta_h) \qquad \begin{pmatrix} h = 1,2,\ldots,m \\ k = 1,2,\ldots,n \end{pmatrix}$$

*are algebraically independent over* $C(z)$ .

Since the functions (9) evidently are Siegel $E$-functions, it follows then from Theorem 10 that *the function values*

$$f(\omega_k;\beta_0),\ f(\Omega_k;\beta_h) \qquad \begin{pmatrix} h = 1,2,\ldots,m \\ k = 1,2,\ldots,n \end{pmatrix}$$

*are algebraically independent over* $Q$ , a result containing the theorem of Hermite-Lindemann.

We shall not prove Shidlovski's result (8) and refer to his original paper. Instead we shall study a similar problem connected with the partial derivatives $\left(\frac{\partial}{\partial\beta}\right)^{k}f(z;\beta)$ of $f(z;\beta)$ . Before doing so, it is convenient first to establish a lemma by Shidlovski which allows a reduction of inhomogeneous problems of algebraic independence to homogeneous ones. In §100 this idea of Shidlovski was already referred to; in special cases, it was already used by Siegel.

**103.** For this purpose, let $L$ be any field of analytic functions of $z$ which is closed under differentiation. Let $m \geq 2$ ; let

$$Q^{*} : w_h' = q_{h0} + \sum_{k=1}^{m} q_{hk}w_k \qquad (h = 1,2,\ldots,m)$$

be an arbitrary inhomogeneous system of linear differential equations with coefficients in $L$ , and let further

$$Q : w'_h = \sum_{k=1}^{m} q_{hk} w_k \qquad (h = 1,2,\ldots,m)$$

be the corresponding homogeneous system.

Assume that $Q^*$ admits a special solution $\mathbf{f}(z)$ with analytic components $f_1(z), \ldots, f_m(z)$ such that $f_1(z), \ldots, f_{m-1}(z)$ , say, are algebraically independent over $L$ , but $f_1(z), \ldots, f_m(z)$ are algebraically dependent over $L$ . It follows that there exists an irreducible polynomial $P(y_1,\ldots,y_m) \not\equiv 0$ in $L[y_1,\ldots,y_m]$ , where $y_1, \ldots, y_m$ are indeterminates over $L$ , such that

(10): $$P\big(f_1(z),\ldots,f_m(z)\big) \equiv 0 .$$

As a polynomial over $L$ , the coefficients in $P$ are analytic functions of $z$ ; differentiating them yields the function $\partial P/\partial z$ . Put

(11): $$\Pi(y_1,\ldots,y_m) = \frac{\partial P}{\partial z} + \sum_{h=1}^{m} \frac{\partial P}{\partial y_h} \left(q_{h0} + \sum_{k=1}^{m} q_{hk} y_k\right) ,$$

so that also $\Pi$ is a polynomial in $L[y_1,\ldots,y_m]$ , and denote further by $P_H(y_1,\ldots,y_m)$ and by $\Pi_H(y_1,\ldots,y_m)$ the highest homogeneous parts of $P$ and of $\Pi$ , respectively; thus $P_H$ is the sum of all terms $p_{(h)}(z) y_1^{h_1} \ldots y_m^{h_m}$ of $P$ which have maximum total degree $h_1 + \ldots + h_m = \nabla(P)$ , and similarly for $\Pi_H$ .

It follows from these definitions that

(12): $$\Pi_H(y_1,\ldots,y_m) = \frac{\partial}{\partial z} P_H + \sum_{h=1}^{m} \left(\frac{\partial}{\partial y_h} P_H\right)\left(\sum_{k=1}^{m} q_{hk} y_k\right) ;$$

(13): $\frac{d}{dz} P(w_1,\ldots,w_m) = \Pi(w_1,\ldots,w_m)$ if $\mathbf{w} = \mathbf{w}(z)$ is a solution of $Q^*$ ;

(14): $\frac{d}{dz} P_H(w_1,\ldots,w_m) = \Pi_H(w_1,\ldots,w_m)$ if $\mathbf{w}$ is a solution of $Q$ .

From (10) and (13),

(15): $$\Pi\big(f_1(z),\ldots,f_m(z)\big) \equiv 0 .$$

Since $P$ is irreducible, this identity requires that $\Pi$ be divisible by $P$ . For otherwise it would be possible to eliminate the function $f_m(z)$ from (10) and (15), and thus $f_1(z), \ldots, f_{m-1}(z)$ would be algebraically dependent over $L$ , contrary to the hypothesis.

Hence there exists an element $A(z)$ in $L$ such that

$$\Pi(y_1,\ldots,y_m) = A(z)P(y_1,\ldots,y_m)$$

identically in $y_1, \ldots, y_m$ . On taking here on both sides only the terms of highest total degree, it follows that similarly

$$\Pi_H(y_1,\ldots,y_m) = A(z)P_H(y_1,\ldots,y_m)$$

identically in $y_1, \ldots, y_m$ . On combining this second identity with (14), we find that

$$\frac{d}{dz} P_H(w_1,\ldots,w_m) = A(z)P_H(w_1,\ldots,w_m) \quad \text{if } \mathbf{w} \text{ is a solution of } Q ,$$

and hence that under this condition

$$P_H(w_1,\ldots,w_m) = ce^{\int A(z)dz} ,$$

where $c$ , the constant of integration, depends on the special solution $\mathbf{w}$ of $Q$ .

We are assuming that $m \geq 2$ . There exist then at least two independent solutions of $Q$ , $\mathbf{u}$ with the components $u_1, \ldots, u_m$ , and $\mathbf{v}$ with the components $v_1, \ldots, v_m$ , say. If $\lambda$ and $\mu$ are two arbitrary constants, also $\lambda\mathbf{u} + \mu\mathbf{v}$ is a solution of $Q$ , and hence

$$P_H(\lambda u_1+\mu v_1,\ldots,\lambda u_m+\mu v_m) = c(\lambda,\mu)e^{\int A(z)dz}$$

where $c(\lambda,\mu)$ evidently is a binary form in $\lambda$ and $\mu$ of degree $\nabla(P_H) = \nabla(P) > 0$ . It is therefore possible to choose two numbers $\lambda_0$ and $\mu_0$ , not both zero, such that $c(\lambda_0,\mu_0) = 0$ . On putting $\mathbf{g}(z) = \lambda_0\mathbf{u} + \mu_0\mathbf{v}$ , the following result is obtained.

(16): *Let $m \geq 2$ . Assume the inhomogeneous system $Q^*$ has a solution $\mathbf{f}(z)$ where certain $m - 1$ of the components of $\mathbf{f}(z)$ are algebraically independent over $L$ , while all $m$ components are algebraically dependent over $L$ . Then there exists a solution $\mathbf{g}(z) \not\equiv \mathbf{0}$ of the corresponding homogeneous system $Q$ , the components of which are algebraically H-dependent over $L$ .*

On combining this lemma with the considerations in §92 of the preceding chapter, one immediately obtains the following analogous result for the solutions of linear differential equations of higher order.

(17): *Let $m \geq 2$ . Assume the linear inhomogeneous differential equation*

$$w^{(m)} + p_1 w^{(m-1)} + \ldots + p_m w = p_0$$

*with coefficients in $L$ has a solution $f(z)$ such that*

*$f(z), f'(z), \ldots, f^{(m-2)}(z)$ are algebraically independent over $L$, but that $f(z), f'(z), \ldots, f^{(m-1)}(z)$ are algebraically dependent over $L$. Then the linear homogeneous differential equation*

$$w^{(m)} + p_1 w^{(m-1)} + \ldots + p_m w = 0$$

*has a solution $g(z) \not\equiv 0$ such that $g(z), g'(z), \ldots, g^{(m-1)}(z)$ are algebraically H-dependent over $L$.*

In the lowest case, when $m = 2$, this lemma implies that the logarithmic derivative $g'(z)/g(z)$ is a function algebraic over $L$.

**104.** We return now to the study of $f(z;\beta)$, but make a slight change of notation. Denote now by $\beta$ a number distinct from $-1, -2, -3, \ldots$, and put

$$f_0(z;\beta) = \sum_{n=0}^{\infty} \frac{1}{(\beta+1)(\beta+2)\ldots(\beta+n)} z^n .$$

Then $f_0(z;\beta)$ satisfies the linear differential equation

$$w' = \frac{\beta}{z} + \left(1 - \frac{\beta}{z}\right)w ,$$

and for $\beta > -1$ it allows the integral representation

(18):
$$f_0(z;\beta) = 1 + z^{-\beta}e^{z} \int_0^z z^{\beta}e^{-z}dz .$$

In addition to $f_0(z;\beta)$ we also consider *all its successive partial derivatives*

$$f_k(z;\beta) = \left(\frac{\partial}{\partial\beta}\right)^k f_0(z;\beta) \qquad (k = 1,2,3,\ldots) .$$

It is obvious that, for every integer $m \geq 0$, the $m + 1$ functions

$$f_k(z;\beta) \qquad (k = 0,1,\ldots,m)$$

satisfy the following inhomogeneous system of linear differential equations,

$$Q_m^* : \begin{cases} w_0' = \frac{\beta}{z} + \left(1 - \frac{\beta}{z}\right)w_0 , \\ w_1' = \frac{1}{z} - \frac{1}{z}w_0 + \left(1 - \frac{\beta}{z}\right)w_1 , \\ w_k' = -\frac{k}{z}w_{k-1} + \left(1 - \frac{\beta}{z}\right)w_k \qquad (k = 2,3,\ldots,m) . \end{cases}$$

If $\beta$ is a rational number, say $\beta = c/d$ where the integers $c$ and $d$ satisfy $(c,d) = 1$ and $d \geq 1$, *all these functions $f_k(z;\beta)$ are again E-functions.* For put

$$\left(\frac{d}{d\beta}\right)^k\{(\beta+1)(\beta+2)\ldots(\beta+n)\}^{-1} = \{(\beta+1)(\beta+2)\ldots(\beta+n)\}^{-1}p_k(\beta;n) .$$

Then

$$p_0(\beta;n) = 1 , \quad p_1(\beta;n) = -\left(\frac{1}{\beta+1} + \frac{1}{\beta+2} + \ldots + \frac{1}{\beta+n}\right) ,$$

and generally

$$p_{k+1}(\beta;n) = p_1(\beta;n)p_k(\beta;n) + \frac{d}{d\beta}\, p_k(\beta;n) .$$

On applying this recursive formula repeatedly, it follows that $p_k(\beta;n)$ can be expressed as a polynomial with integral coefficients in the expressions

$$\sum_{j=1}^{n} (c+dj)^{-h} \qquad (h = 1,2,\ldots,k) .$$

Here, by §96, the least common multiple of the integers

$$c+d,\ c+2d,\ \ldots,\ c+dn$$

in the denominators is for large $n$ of the order

$$O(C^n)$$

where $C > 1$ is a constant independent of $n$. Since $f_k(z;\beta)$, for $k \geq 1$, has the explicit form

$$f_k(z;\beta) = \sum_{n=1}^{\infty} \frac{p_k(\beta;n)}{(\beta+1)(\beta+2)\ldots(\beta+n)}\, z^n ,$$

it follows easily from this estimate and from the earlier estimate (1) that $f_k(z;\beta)$ is an $E$-function relative to $\mathbf{Q}$, and, as a solution of $Q_m^*$, it is a Siegel $E$-function.

**105.** We shall now prove the following general result.

**(19):** *Let* $w_0 = g_0(z),\ w_1 = g_1(z),\ \ldots$ *be a solution of the system of linear differential equations*

$$w_0' = q_{00}w_0 , \quad w_h' = q_h + \sum_{k=0}^{h} q_{hk}w_k \qquad (h = 1,2,\ldots) ,$$

*where the coefficients* $q_h,\ q_{hk}$ *are rational functions of* $z$. *Suppose further that*

(1) $g_0(z)$ *is a transcendental function;*

(2) *the differential equation* $w' + \left(\sum_{k=0}^{m} s_k q_{kk}\right)w = 0$ *has only the trivial rational solution if* $m \geq 1$ *and the* $s_k$ $(k = 0,1,\ldots,m)$ *are non-negative integers, with* $s_0 \geq 1$ , $s_m \geq 1$ ;

(3) *the differential equation* $w' = q_{11}w - q_1$ *has no rational solution;*

(4) *the differential equation* $w' = -\left(\sum_{k=1}^{M-1} u_k q_{kk} - q_{MM}\right)w$ *has only the trivial rational solution if* $M \geq 2$ *and the* $u_k$ $(k = 1,2,\ldots,M-1)$ *are non-negative integers satisfying* $u_{M-1} \geq 1$ *and* $\sum_{k=1}^{M-1} u_k \geq 2$ , *and finally*

(5) *the differential equation* $w' = (q_{MM}-q_{M-1,M-1})w - q_{M,M-1}$ *has no rational solution if* $M \geq 2$ .

*Then any finite number of the functions* $g_0, g_1, g_2, \ldots$ *are algebraically independent over* $\mathbb{C}(z)$ .

Let us assume that the statement is false. Then there exists a (unique) suffix $m$ such that the functions

$$g_0, g_1, \ldots, g_{m-1}$$

are algebraically independent, while the functions

$$g_0, g_1, \ldots, g_m$$

are algebraically dependent, over $\mathbb{C}(z)$ . Because of the assumption (1), we have $m \geq 1$ . This means that there exists a non-constant irreducible polynomial $P(w_0,w_1,\ldots,w_m)$ over $\mathbb{C}(z)$ such that $P(g_0,g_1,\ldots,g_m) \equiv 0$ . The degree of $P$ with respect to the variable $w_m$ must be at least 1 . If $\Pi(w_0,w_1,\ldots,w_m)$ is another polynomial over $\mathbb{C}(z)$ with $\Pi(g_0,g_1,\ldots,g_m) \equiv 0$ , then there exists a third polynomial $p(w_0,w_1,\ldots,w_m)$ over $\mathbb{C}(z)$ such that $\Pi = pP$ . In particular, if the total degree of $\Pi$ is equal to that of $P$ , then $p \in \mathbb{C}(z)$ . If the total degree of $\Pi$ is less than that of $P$ , then $p \equiv 0$ .

Let us now define

(20): $$\Pi(w_0,w_1,\dots,w_m) = \frac{\partial P}{\partial z} + \frac{\partial P}{\partial w_0} q_{00}w_0 + \sum_{k=1}^{m} \frac{\partial P}{\partial w_k}(q_k+q_{k0}w_0+q_{k1}w_1+\dots+q_{kk}w_k) =$$

$$= \frac{\partial P}{\partial z} + \sum_{k=0}^{m} \frac{\partial P}{\partial w_k} q_{kk}w_k + \sum_{k=1}^{m} \frac{\partial P}{\partial w_k} q_{k,k-1}w_{k-1} + \dots + \frac{\partial P}{\partial w_m} q_{m0}w_0 + \sum_{k=1}^{m} \frac{\partial P}{\partial w_k} q_k .$$

Since $\Pi\big(g_0(z),g_1(z),\dots,g_m(z)\big) = \frac{d}{dz} P\big(g_0(z),g_1(z),\dots,g_m(z)\big)$ , we have $\Pi(g_0,g_1,\dots,g_m) \equiv 0$ and since the total degree of $\Pi$ does not exceed that of $P$ , we have $\Pi = pP$ , where $p \in C(z)$ . Let us denote the sum of those terms of $P$ (or $\Pi$ ) in which the variable $w_0$ does not occur by $P_0$ (or $\Pi_0$ ). There must be at least one non-zero term in $P_0$ , since otherwise $P$ would be divisible by $w_0$ and thus reducible ($P = w_0$ is excluded, since $g_0 \not\equiv 0$ ). From the relation (20) it is clear that terms of the polynomial $\Pi_0$ can come only from the terms of the polynomial $P_0$ . For when we go over from $P$ to $\Pi$ , the exponent on the variable $w_0$ never decreases. Thus we even have

(21): $$\Pi_0 = pP_0 .$$

In the following, we shall write for simplicity $\{i_1,i_2,\dots,i_m\}$ instead of $w_1^{i_1} w_2^{i_2} \dots w_m^{i_m}$ . Let $R_{(i)}(z)$ and $R_{(i')}(z)$ be two rational functions (possibly $\equiv 0$ ). We shall say that the term $R_{(i)}.\{i_1,i_2,\dots,i_m\}$ is higher (lower) than $R_{(i')}.\{i'_1,i'_2,\dots,i'_m\}$ , if the first non-zero difference $i_m-i'_m$, $i_{m-1}-i'_{m-1}$, ..., $i_1-i'_1$ is positive (negative).

Without loss of generality, we may suppose that the coefficient of the highest non-zero term of $P_0$ is equal to 1 . We can then write

$$P_0 = \{t_1,t_2,\dots,t_m\} + \text{lower terms.}$$

Let $M$ be the smallest suffix $k$ for which $t_k \ge 1$ . First we observe that necessarily $M \le m$ , since otherwise we would have $P_0 \equiv 1$ , whence $\Pi_0 \equiv 0$ , $p \equiv 0$ and $\Pi \equiv 0$ . But $\Pi \equiv 0$ is impossible. For let

$$P = R_{(s)}.\{s_0,s_1,\dots,s_m\} + \text{lower terms,}$$

where $R_{(s)} \not\equiv 0$ . Then we must have $s_0 \ge 1$ , $s_m \ge 1$ and

$$\Pi = \left[R'_{(s)} + \left(\sum_{k=0}^{m} s_k q_{kk}\right) R_{(s)}\right]\{s_0,s_1,\dots,s_m\} + \text{lower terms.}$$

If $\Pi$ were identically $0$ , we should have $R'_{(s)} + \left(\sum_{k=0}^{m} s_k q_{kk}\right) R_s = 0$ . However, by by the assumption (2), this differential equation does not have any nontrivial rational solution. It follows that $1 \leq M \leq m$ .

First suppose that $M = 1$ . In this case we have

$$P_0 = \{t_1, t_2, \dots, t_m\} + R\{t_1 - 1, t_2, \dots, t_m\} + \text{lower terms},$$

$$\Pi_0 = \left[R' + \left(\sum_{k=1}^{m} t_k q_{kk} - q_{11}\right) R + t_1 q_1\right] \{t_1 - 1, t_2, \dots, t_m\} + \left(\sum_{k=1}^{m} t_k q_{kk}\right) \{t_1, t_2, \dots, t_m\} + \\ + \text{lower terms}.$$

By (21), we have

$$p = \sum_{k=1}^{m} t_k q_{kk} \ , \quad pR = R' + \left(\sum_{k=1}^{m} t_k q_{kk} - q_{11}\right) R + t_1 q_1 \ ,$$

whence $R' = q_{11} R - t_1 q_1$ . Put here $R = t_1 w$ . Using the assumption (3), we see that this differential equation does not have any rational solution and so $M \geq 2$ .

Thus $P_0$ is of the form

$$P_0 = \{0, \dots, 0, t_M, \dots, t_m\} + R\{u_1, \dots, u_{M-1}, t_M - 1, t_{M+1}, \dots, t_m\} + \text{lower terms},$$

where $u_{M-1} \geq 1$ . Suppose for the moment that $\sum_{k=1}^{M-1} u_k \geq 2$ . Under these conditions, we have

$$\Pi_0 = \left(\sum_{k=M}^{m} t_k q_{kk}\right) \{0, \dots, 0, t_M, \dots, t_m\} + \left[R' + \left(\sum_{k=1}^{M-1} u_k q_{kk} + \sum_{k=M}^{m} t_k q_{kk} - q_{MM}\right) R\right] \times \\ \times \{u_1, \dots, u_{M-1}, t_M - 1, t_{M+1}, \dots, t_m\} + \text{lower terms}.$$

Thus $p = \sum_{k=M}^{m} t_k q_{kk}$ , $pR = R' + \left(\sum_{k=1}^{M-1} u_k q_{kk} + \sum_{k=M}^{m} t_k q_{kk} - q_{MM}\right) R$ and $R' = -\left(\sum_{k=1}^{M-1} u_k q_{kk} - q_{MM}\right) R$ . By the assumption (4), this last equation has only the trivial rational solution $R \equiv 0$ .

It follows that we may write

$$P_0 = \{0, \dots, 0, t_M, \dots, t_m\} + R\{0, \dots, 0, 1, t_M - 1, t_{M+1}, \dots, t_m\} + \text{lower terms}.$$

Then

$$\Pi_0 = \left(\sum_{k=M}^{m} t_k q_{kk}\right)\{0,\ldots,0,t_M,\ldots,t_m\} + \left[R'+\left(\sum_{k=M}^{m} t_k q_{kk}+q_{M-1,M-1}-q_{MM}\right)R+t_M q_{M,M-1}\right] \times$$
$$\times \{0,\ldots,0,1,t_M-1,t_{M+1},\ldots,t_m\} + \text{lower terms},$$

and thus

$$p = \sum_{k=M}^{m} t_k q_{kk} \ , \quad pR = R' + \left(\sum_{k=M}^{m} t_k q_{kk}+q_{M-1,M-1}-q_{MM}\right)R + t_M q_{M,M-1} \ .$$

On eliminating $p$, we get $R' = (q_{MM}-q_{M-1,M-1})R - t_M q_{M,M-1}$. Put here $R = t_M w$. Using the assumption (5), we see that this differential equation does not have any rational solution. This finishes the proof of Lemma (19).

**106.** We are now ready to prove the following pair of results concerning the functions $f_k(z;\beta)$ introduced in §104.

**(22):** *If* $\beta$ *is not a negative integer, then any finite number of the functions*

$$f_k(z;\beta) \qquad (k = 0,1,2,\ldots)$$

*are algebraically independent over* $\mathbb{C}(z)$ .

**(23):** *If* $\beta$ *is not an integer, then any finite number of the functions*

$$e^z \ \text{and} \ f_k(z;\beta) \qquad (k = 0,1,2,\ldots)$$

*are algebraically independent over* $\mathbb{C}(z)$ .

Proof of (23). Let us consider the system of differential equations

$$(24): \quad \begin{cases} w_0' = w_0 \ , \\ w_1' = \beta z^{-1} + (1-\beta z^{-1})w_1 \ , \\ w_2' = z^{-1} - z^{-1}w_1 + (1-\beta z^{-1})w_2 \ , \\ w_h' = -(h-1)z^{-1}w_{h-1} + (1-\beta z^{-1})w_h \qquad (h = 3,4,\ldots) \ . \end{cases}$$

This system has solution $w_0 = e^z$, $w_h = f_{h-1}$ $(h = 1,2,\ldots)$. In order to prove (23), it suffices to verify that all the assumptions of Lemma 19 are satisfied. We have

$$q_{00} = 1 \ , \quad q_{kk} = 1 - \beta z^{-1} \qquad (k = 1,2,\ldots) \ ,$$

$$q_1 = \beta z^{-1} \ , \quad q_{k,k-1} = -(k-1)z^{-1} \qquad (k = 2,3,\ldots) \ .$$

(1) $g_0(z) = e^z$ is a transcendental function.

(2) The only rational solution of the differential equation $w' + \left\{s_0+\left(\sum_{k=1}^{m} s_k\right)(1-\beta z^{-1})\right\}w = 0$ is $w \equiv 0$ , provided that $m \geq 1$ , $s_0 \geq 1$ , $s_m \geq 1$ , and $s_k \geq 0$ for $k = 0, 1, \ldots, m$ . For, on integrating the equation directly, we get $w = \text{const.}z^{\beta(\sigma-s_0)}e^{-\sigma z}$ where $\sigma = \sum_{k=0}^{m} s_k \geq 2$ .

(3) The differential equation $w' = (1-\beta z^{-1})w - \beta z^{-1}$ does not have any rational solution when $\beta$ is not an integer. To see this, note first that from the form of the differential equation, $w(z)$ can have a pole in the finite plane only at $z = 0$ . If there is such a pole, of order $l \geq 1$ , so that

$$s(z) = c_0 z^{-l} + c_1 z^{-l+1} + \ldots , \text{ where } c_0 \neq 0 ,$$

we find, on comparing the two sides of the differential equation, that

$$-lc_0 z^{-l-1} + \ldots = -\beta c_0 z^{-l-1} + \ldots ,$$

whence $\beta = l$ , an integer, contrary to hypothesis. Thus $w(z)$ must be a polynomial. But this too is impossible, for the degree of $w'$ would be less than that of the right-hand side in the differential equation.

(4) The differential equation $w' = -\left(\sum_{k=1}^{M-1} u_k - 1\right)(1-\beta z^{-1})w$ , where $\sum_{k=1}^{M-1} u_k - 1 \geq 1$ , has only the trivial rational solution. This is again clear from the form of the solution: $w = \text{const.}z^{\beta(\sigma-1)}e^{(1-\sigma)z}$ where $\sigma = \sum_{k=1}^{M-1} u_k$ .

(5) The differential equation $w' = (M-1)z^{-1}$ does not have any rational solution, provided $M \geq 2$ .

This concludes the proof of (23).

Proof of (22). Since we have already proved (23), we may restrict ourselves in the proof of (22) to the case when $\beta$ is a non-negative integer. Because we want to use Lemma (19) we have to change the system satisfied by $f_0, f_1, \ldots$ slightly. We shall put $w_0 = \overline{w}_0 - \beta! \sum_{k=1}^{\beta} \frac{z^{-k}}{(\beta-k)!}$ . After simple calculations, we obtain the new system of differential equations

$$\bar{w}_0' = (1-\beta z^{-1})\bar{w}_0 ,$$

$$w_1' = \beta! z^{-(\beta+1)} \sum_{k=0}^{\beta} \frac{z^k}{k!} - z^{-1}\bar{w}_0 + (1-\beta z^{-1})w_1 ,$$

$$w_h' = -hz^{-1}w_{h-1} + (1-\beta z^{-1})w_h \qquad (h = 2,3,\ldots) ,$$

which has the form required. This system has the solution

$$\bar{w}_0 = f_0 + \beta! \sum_{k=1}^{\beta} \frac{z^{-k}}{(\beta-k)!} = \bar{f}_0 ,$$

while $w_h = f_h$ for $h = 1, 2, \ldots$ . If the functions $\bar{f}_0, f_1, \ldots, f_m$ are algebraically independent, then also the functions $f_0, f_1, \ldots, f_m$ are algebraically independent, and vice versa. Thus it again suffices to verify that all the conditions of Lemma (19) are satisfied.

(1) $\bar{f}_0$ is a transcendental function, since $f_0$ is .

(2) The only rational solution of the differential equation $w' + \left(\sum_{k=0}^{m} s_k\right)(1-\beta z^{-1})w = 0$ , where $\sum_{k=0}^{m} s_k \geq 2$ , is $w = 0$ . This follows from the direct solution again: $w = \text{const.}e^{-\sigma z}z^{\beta\sigma}$ , where $\sigma = \sum_{k=0}^{m} s_k$ .

(3) The differential equation $w' = (1-\beta z^{-1})w - \beta! z^{-(\beta+1)} \sum_{k=0}^{\beta} \frac{z^k}{k!}$ does not have any rational solution. From the differential equation, the only possible finite singularity is $z = 0$ . If $R$ were a rational solution with a pole at zero, we would have, in a neighbourhood of $0$ , $R = a_k z^{-k}$ + higher terms, where $k \geq 1$ , $a_k \neq 0$ . Thus $R' = -ka_k z^{-(k+1)}$ + higher terms, while

$$(1-\beta z^{-1})R - \beta! z^{-(\beta+1)} \sum_{k=0}^{\beta} \frac{z^k}{k!} = \left(-\beta a_k z^{-(k+1)} + \text{higher terms}\right) + \\ + \left(-\beta! z^{-(\beta+1)} + \text{higher terms}\right).$$

For $k > \beta$ we have $-ka_k = -\beta a_k$ , so $k = \beta$ , a contradiction. Both of the cases $k = \beta$ and $k < \beta$ lead to $-\beta! = 0$ , and thus to a contradiction. Hence $R$ cannot have any finite singularity and thus, as a rational function, it must be a polynomial.

But this is also impossible. For $R \equiv 0$ is clearly not a solution, while if

$R \not\equiv 0$ , then $R = b_l z^l$ + lower terms, where $l \geq 0$ , $b_l \neq 0$ , and $R' = l b_l z^{l-1}$ + lower terms, while on the right-hand side we have $b_l z^l$ + lower terms.

(4) The differential equation $w' = -\left(\sum_{k=1}^{M-1} u_k - 1\right)(1-\beta z^{-1})w$ , where $\sum_{k=1}^{M-1} u_k \geq 2$ , has only the trivial rational solution, as in (2).

(5) The differential equation $w' = Mz^{-1}$ , where $M \geq 2$ , does not have any rational solution.

This finishes the proof of (22).

REMARK. Lemma (19) was formulated exactly to make the proofs of (22) and (23) as simple as possible. It would, of course, be possible to prove a more general theorem applicable to other systems of triangular type.

107. The functions $f_k(z)$ and $e^z$ considered in the Lemmas (22) and (23) are Siegel $E$-functions; and the coefficients in the system (24) have the least common denominator $z$ . We can therefore immediately apply Theorem 10 and arrive at the following two results.

**(25):** *If* $\alpha \neq 0$ *is an algebraic number, and* $\beta$ *is a rational number which is not a negative integer, every finite number of function values*

$$f_k(\alpha;\beta) \qquad (k = 0,1,2,\ldots)$$

*are algebraically independent over* $\mathbb{Q}$ .

**(26):** *If* $\alpha \neq 0$ *is an algebraic number, and* $\beta$ *is a rational number which is not an integer, every finite number of function values*

$$e^\alpha,\ f_k(\alpha;\beta) \qquad (k = 0,1,2,\ldots)$$

*are algebraically independent over* $\mathbb{Q}$ .

Thus in both lemmas all the function values are transcendental. By way of example, the two series

$$f_0(\alpha;0) = e^\alpha = \sum_{n=0}^{\infty} \frac{\alpha^n}{n!} \quad \text{and} \quad f_1(\alpha;0) = -\sum_{n=1}^{\infty} \left(\frac{1}{1} + \frac{1}{2} + \ldots + \frac{1}{n}\right) \frac{\alpha^n}{n!}$$

are algebraically independent over $\mathbb{Q}$ if $\alpha \neq 0$ is algebraic.

The results (25) and (26) may be applied to the integrals for $f_k(\alpha;\beta)$ and assume then an even more striking form. By formula (18),

$$f_0(z;\beta) = 1 + z^{-\beta}e^{z} \int_0^z z^{\beta}e^{-z}dz \quad \text{if} \quad \beta > -1 \;.$$

This formula is equivalent to

$$f_0(z;\beta) = 1 + ze^{z} \int_0^1 s^{\beta}e^{-sz}ds \;,$$

and this implies that

$$f_k(z;\beta) = \left(\frac{\partial}{\partial\beta}\right)^k f_0(z;\beta) = ze^{z} \int_0^1 s^{\beta}(\log s)^k e^{-sz}ds \qquad (k = 1,2,3,\ldots) \;.$$

We claim now the following result.

**(27)**: *If* $\alpha \neq 0$ *is an algebraic number, and* $\beta > -1$ *is a rational number which is not an integer, any finite number of the integrals*

$$\int_0^1 s^{\beta}(\log s)^k e^{-\alpha s}ds \qquad (k = 0,1,2,\ldots)$$

*are algebraically independent over* $Q$ *.*

For let us put $T_k = \int_0^1 s^{\beta}(\log s)^k e^{-\alpha s}ds \quad (k = 0,1,2,\ldots)$ . If $\beta > -1$ is a

$f_{k_1}(\alpha,\beta),\ f_{k_2}(\alpha,\beta),\ \ldots,\ f_{k_n}(\alpha,\beta)$ is algebraically independent over $Q$ , for $\alpha \neq 0$ and algebraic. This implies that the system

$$\alpha e^{\alpha}T_{k_1},\ \alpha e^{\alpha}T_{k_2},\ \ldots,\ \alpha e^{\alpha}T_{k_n} \qquad (k_1 < k_2 < \ldots < k_n)$$

is algebraically independent over $Q$ .

This, in turn, implies that the system

$$e^{\alpha}T_{k_1},\ e^{\alpha}T_{k_2},\ \ldots,\ e^{\alpha}T_{k_n} \qquad (k_1 < k_2 < \ldots < k_n)$$

is algebraically independent over $Q$ ($\alpha$ is algebraic and we may proceed by the standard resultant elimination technique).

If $\beta > -1$ is a rational number which is not an integer, then (see (26)), using the same reasoning, we get the algebraic independence over $Q$ of the system

$$e^{\alpha},\ e^{\alpha}T_{k_1},\ e^{\alpha}T_{k_2},\ \ldots,\ e^{\alpha}T_{k_n} \qquad (k_1 < k_2 < \ldots < k_n) \;.$$

This already yields the algebraic independence of the system

$$T_{k_1}, T_{k_2}, \ldots, T_{k_n} \qquad (k_1 < k_2 < \ldots < k_n) ,$$

as claimed in (27). For, the algebraic dependence of the second system implies that of the first (just write "$T = e^{-\alpha}.e^{\alpha}T$ ").

108. Let us now proceed to the study of $E$-functions that satisfy either a linear differential equation

$$(28): \qquad w^{(m)} + p_1 w^{(m-1)} + \ldots + p_m w = p_0$$

of order $m \geq 2$ and with coefficients in $\mathbb{C}(z)$ , or an equivalent system of linear differential equations of the first order. The question which needs to be answered is whether a solution $w$ of (28) can in addition satisfy an algebraic differential equation

$$(29): \qquad P(w,w',\ldots,w^{(M)}) = 0$$

of order $M$ where $0 \leq M \leq m-1$ (which for $M = 0$ becomes an algebraic equation) and where further $P(w_0,w_1,\ldots,w_M) \not\equiv 0$ denotes a polynomial in $w_0, w_1, \ldots, w_M$ with coefficients in $\mathbb{C}(z)$ .

In the paper already referred to in §100, Oleinikov proposed the following simple method for solving this problem, and he showed in a number of examples that his method is quite effective for small values of $m$ .

Since $M$ is an arbitrary integer in $0 \leq M \leq m-1$ , we may assume that

$$\frac{\partial P}{\partial w_M} \not\equiv 0 .$$

Then the equation (29) defines $w^{(M)}$ as an algebraic function

$$w^{(M)} = A(z,w,w',\ldots,w^{(M-1)})$$

of $z, w, w', \ldots, w^{(M-1)}$ . By the theory of algebraic functions it is therefore possible to develop $w^{(M)}$ , for example, into a *descending* Laurent series

$$(30): \qquad w^{(M)} = \sum_{n=0}^{\infty} a_n(z,w,w',\ldots,w^{(M-2)})(w^{(M-1)})^{e_n}$$

where the exponents $e_n$ are rational numbers, all with the same denominator, such that

$$e_0 > e_1 > e_2 > \ldots ,$$

and where the coefficients $a_n$ are algebraic functions of $z, w, w', \ldots, w^{(M-2)}$ .

There are also analogous series for $w^{(M)}$ in ascending powers of $w^{(M-1)} - c$ with exponents $e_n$ and coefficients $a_n$ of the same kind, but where now

$$e_0 < e_1 < e_2 < \dots$$

and where $c$ is an arbitrary constant. We are not concerned with the convergence of these series (which can actually be established for all values of $w^{(M-1)}$ sufficiently near to $\infty$ or to $c$, respectively), but rather with their formal behaviour.

In order to decide whether a solution of (28) can satisfy some algebraic differential equation (29) of a certain order $M$, one postulates for $w^{(M)}$ the development (30), say, substitutes in (28), and tries to choose the exponents $e_n$ and the coefficients $a_n$ such that all terms cancel out. If this proves impossible, the question has a negative answer for the order $M$. Otherwise one obtains properties of the algebraic function $A$ and hence of the proposed equation (29), and this may allow one to establish the conditions for the parameters under which such an equation can hold.

We shall apply this method in the next sections to a number of examples. In particular, we use this method to establish Siegel's main theorem on Bessel functions.

**109.** In the special case when $r = 0$, $s = 2$, $t = 2$ the hypergeometric $E$-function of §98 takes the form

$$f(z;\beta_1,\beta_2) = \sum_{n=0}^{\infty} \frac{1}{[\beta_1,n][\beta_2,n]} (z/2)^{2n} ,$$

and it satisfies the linear differential equation

$$\left((\theta+\beta_1-1)(\theta+\beta_2-1)-(z/2)^2\right)w = (\beta_1-1)(\beta_2-1) ,$$

which may be written in the explicit form

$$w'' + \frac{2\beta_1+2\beta_2-3}{z} w' + \left(\frac{4(\beta_1-1)(\beta_2-1)}{z^2} - 1\right) w = \frac{4(\beta_1-1)(\beta_2-1)}{z^2} .$$

Here it is convenient to change the notation slightly. Put

$$\beta_1 = \lambda + 1 , \quad \beta_2 = \mu + 1 ,$$

and replace $z$ by $iz$. This change leads to the new function

(31): $$K(z;\lambda,\mu) = \sum_{n=0}^{\infty} \frac{(-z^2/4)^n}{[\lambda+1,n][\mu+1,n]}$$

which satisfies the differential equation

(32): $$w'' + \frac{2\lambda+2\mu+1}{z} w' + \left(1 + \frac{4\lambda\mu}{z^2}\right) w = \frac{4\lambda\mu}{z^2} .$$

In the special case when $\lambda = \nu$ , $\mu = 0$ , this function is connected with the ordinary Bessel function of the first kind by the identity

$$J_\nu(z) = \frac{(z/2)^\nu}{\Gamma(\nu+1)} K(z;\nu,0) .$$

By §99, $K(z;\lambda,\mu)$ is a Siegel $E$-function whenever $\lambda$ and $\mu$ are rational numbers distinct from $-1, -2, -3, \ldots$ .

**110.** The problem is to find conditions on $\lambda$ and $\mu$ under which no solution $w$ of the differential equation (32) either is an algebraic function not identically zero, or satisfies an algebraic differential equation of the first order.

Assume, firstly, that $w \not\equiv 0$ is an algebraic function. By the theory of algebraic functions, in a neighbourhood of $z = \infty$ , $w$ can be written as a series

$$w = \sum_{k=-\infty}^{m} a_k z^{k/n}$$

where $m$ and $n \geq 1$ are integers, and where the highest coefficient $a_m$ does not vanish. Since

$$w' = \sum_{k=-\infty}^{m} a_k \frac{k}{n} z^{(k/n)-1} , \quad w'' = \sum_{k=-\infty}^{m} a_k \frac{k}{n} \left(\frac{k}{n} - 1\right) z^{(k/n)-2} ,$$

the differential equation (32) is easily seen to be equivalent to

(33): $$\sum_{k=-\infty}^{m} a_k \left(\frac{k}{n} + 2\lambda\right)\left(\frac{k}{n} + 2\mu\right) z^{(k/n)-2} + \sum_{k=-\infty}^{m} a_k z^{k/n} = \frac{4\lambda\mu}{z^2} .$$

In this development, all terms have to cancel out. On the left-hand side the term involving the highest power of $z$ is $a_m z^{m/n}$ , which necessarily is identical with the term $\frac{4\lambda\mu}{z^2}$ on the right-hand side; hence

$$\frac{m}{n} = -2 \quad \text{and} \quad a_m = 4\lambda\mu .$$

Therefore, in particular, the assumption $a_m \neq 0$ implies already that

(34): $$\lambda \neq 0 \quad \text{and} \quad \mu \neq 0 .$$

We next assert that the exponents $\frac{k}{n}$ of $z$ are integers so that we may take

$$n = 1 .$$

For otherwise let $M$ be the largest suffix for which

$$a_M \neq 0 \ , \text{ and } \frac{M}{n} \text{ is not an integer.}$$

Then $\frac{M}{n} < -2$ , and so the term

$$a_M z^{M/n}$$

on the left-hand side of (33) cannot be cancelled by any other term in this identity.

The power series for $w$ thus has the simpler form

$$w = \sum_{k=-\infty}^{-2} a_k z^k \ , \text{ where } a_{-2} = 4\lambda\mu \ .$$

This means that

$w$ is regular at $z = \infty$ and has there a zero of order 2 .

By the differential equation (32), then, the only possible singular point of the algebraic function lies at $z = 0$ . On the other hand, any algebraic function which is not a rational function has at least 2 branch points. It follows that

$w$ is a rational function of $z$ with at most one pole, at $z = 0$ .

Hence $w$ can be written in the form

$$w = \frac{p(z)}{z^r}$$

where $p(z) \not\equiv 0$ is a polynomial, and $r \geq 0$ is an integer. Since $w$ has a double zero at $z = \infty$ , it is a polynomial in $z^{-1}$ and so can be written as

$$w = \sum_{k=2}^{s} A_k z^{-k} \ ,$$

where $s \geq 2$ is an integer, and where, in particular, $A_2$ has the value

$$A_2 = 4\lambda\mu \ .$$

Thus, we obtain from (32),

(35): $\displaystyle\sum_{k=2}^{s} A_k(k-2\lambda)(k-2\mu)z^{-(k+2)} + \sum_{k=3}^{s} A_k z^{-k} = 0 \ , \quad s \geq 2 \ , \quad A_s \neq 0 \ , \quad \lambda\mu \neq 0 \ .$

It follows that $(s-2\lambda)(s-2\mu) = 0$ , and by symmetry we may consider the case $s = 2\lambda$ only. The identity (35) yields $A_3 = 0$ . Also, because $A_s \neq 0$ , it implies $A_{s-2} \neq 0, A_{s-4} \neq 0, \ldots$ . Since $A_3 = 0$ , $s$ must be even and $A_3 = A_5 = \ldots = A_{s-1} = 0$ . Hence, $\lambda$ is a natural number $(s = 2\lambda)$ and we may write $w = \sum_{k=1}^{\lambda} A_{2k} z^{-2k}$ . The identity (35) thus becomes

(36): $\sum_{k=1}^{\lambda} A_{2k}(2k-2\lambda)(2k-2\mu)z^{-(2k+2)} + \sum_{k=2}^{\lambda} A_{2k}z^{-2k} \equiv 0$ , $\lambda \geq 1$ , $A_{2\lambda} \neq 0$ , $\mu \neq 0$ .

The identity (36) yields a system of linear equations for the coefficients $A_{2k}$ $(k = 1,2,\ldots,\lambda)$ with the solution

(37): $$A_{2k} = (-1)^{k+1} 2^{2k} (k!)^2 \binom{\lambda}{k} \binom{\mu}{k} \qquad (k = 1,2,\ldots,\lambda) .$$

Since all the coefficients $A_{2k}$ $(k = 1,2,\ldots,\lambda)$ must be different from zero, $\binom{\mu}{k} \neq 0$ for $k = 1, 2, \ldots, \lambda$ or more simply, $\binom{\mu}{\lambda} \neq 0$ . Because of the symmetry between $\lambda$ and $\mu$ , we can state the final result as follows.

*The differential equation* (32) *has no algebraic solution* $w \not\equiv 0$ *except when* $\lambda$ *is a natural number and* $\binom{\mu}{\lambda} \neq 0$ *, or when* $\mu$ *is a natural number and* $\binom{\lambda}{\mu} \neq 0$ *. In the first case* $w = \sum_{k=1}^{\lambda} A_{2k} z^{-2k}$ *, where the* $A_{2k}$ *are given by* (37) *for* $k = 1, 2, \ldots, \lambda$ *; in the other case* $w = \sum_{k=1}^{\mu} A_{2k} z^{-2k}$ *, where the* $A_{2k}$ *are given by* (37) *for* $k = 1, 2, \ldots, \mu$ .

111. There remains the case when the differential equation (32) has an integral which is *not* an *algebraic function*, but which satisfies an *algebraic differential equation of the first order*. By Shidlovski's lemma, this requires that the homogeneous differential equation

$$w'' + \frac{2\lambda+2\mu+1}{z} w' + \left(1 + \frac{4\lambda\mu}{z^2}\right) w = 0$$

have a solution which satisfies a homogeneous algebraic differential equation of the first order. In other words, the quotient

$$W = \frac{w'}{w}$$

must be an algebraic function of $z$ . It is well known, on the other hand, that this quotient satisfies the Riccati equation

(38): $$W' + W^2 + \frac{2\lambda+2\mu+1}{z} W + \left(1 + \frac{4\lambda\mu}{z^2}\right) = 0 .$$

We must thus now decide for which rational values of $\lambda$ and $\mu$ this equation has an algebraic integral, and we shall do so by a method similar to that in §110.

By (38), $W$ is not identically zero. As an algebraic function, it can then again be written in the form

$$W = \sum_{k=-\infty}^{M} A_k z^{k/N}$$

where $M$ and $N > 0$ are integers, and where, since $W \not\equiv 0$,

$$A_M \neq 0 .$$

Therefore, by (38),

$$(39): \quad \left\{\sum_{k=-\infty}^{M} A_k \frac{k}{N} z^{(k/N)-1}\right\} + \left\{\sum_{k=-\infty}^{M} A_k z^{k/N}\right\}^2 + $$
$$+ (2\lambda+2\mu+1) \sum_{k=-\infty}^{M} A_k z^{(k/N)-1} + \left(1 + \frac{4\lambda\mu}{z^2}\right) = 0 .$$

If here $M > 0$, the term

$$A_M^2 z^{2(M/N)}$$

on the left-hand side cannot be cancelled by any other term; and the same is true for the term 1 if $M < 0$.

Therefore necessarily $M = 0$, and thus

$$A_0^2 + 1 = 0 , \quad A_0 = \mp i .$$

Again all the exponents $\frac{k}{N}$ in the series for $W$ must be integers, hence $N = 1$. For let $k < M = 0$ be the largest suffix such that $A_k \neq 0$ and $\frac{k}{N}$ is not integral. Then the corresponding term

$$2A_0 z^0 . A_k z^{k/N}$$

cannot be cancelled by any other term.

This leaves a term

$$(2\lambda+2\mu+1)A_0 z^{-1}$$

on the left-hand side of (39) which can be cancelled only by a term

$$2A_0 z^0 . A_{-1} z^{-1} .$$

Hence

$$A_{-1} = - \frac{2\lambda+2\mu+1}{2} .$$

Thus the series for $W$ begins with

(40): $$W = A_0 - \frac{2\lambda+2\mu+1}{2z} + \dots \qquad (A_0 = \mp i) ,$$

while all the further terms involve negative integral powers of $z$ . This shows that $W$ is regular at $z = \infty$ .

The differential equation for $w$ is regular at every finite point distinct from $z = 0$ ; hence at such a point $W = \frac{w'}{w}$ may have a pole, but cannot have a branch point. The only possible branch point of $W$ is then at $z = 0$ . But again as an algebraic function $W$ cannot have only one branch point. It follows then that $W$ *must be a rational function of* $z$ .

In the neighbourhood of $z = 0$ , $W$ allows a development

$$W = \sum_{k=-n}^{\infty} B_k z^k$$

where $n$ is a certain integer, and $B_{-n}$ does not vanish. Thus, by (38),

$$\sum_{k=-n}^{\infty} B_k k z^{k-1} + \left(\sum_{k=-n}^{\infty} B_k z^k\right)^2 + (2\lambda+2\mu+1) \sum_{k=-n}^{\infty} B_k z^{k-1} + \left(1 + \frac{4\lambda\mu}{z^2}\right) = 0 .$$

Here the terms

$$\frac{4\lambda\mu}{z^2} \text{ for } n \le 0 \text{ , and } B_{-n}^2 z^{-2n} \text{ for } n \ge 2$$

are not cancelled by any other terms, at least if $\lambda\mu \ne 0$ , and hence in this case $n$ has the value $+1$ . In any case, $n \le 1$ and

$$B_{-1}(-1) + B_{-1}^2 + (2\lambda+2\mu+1)B_{-1} + 4\lambda\mu = 0 ,$$

so that either

$$B_{-1} = -2\lambda \text{ or } B_{-1} = -2\mu .$$

Therefore, in the neighbourhood of $z = 0$ , $W$ has one of the two forms

(41): $$W = -\frac{2\lambda}{z} + \sum_{k=0}^{\infty} B_k z^k \text{ or } W = -\frac{2\mu}{z} + \sum_{k=0}^{\infty} B_k z^k .$$

In addition to the simple pole at $z = 0$ , $W$ may have finitely many further poles at certain points

(42): $$z = \gamma \text{ , where } \gamma = \gamma_1, \gamma_2, \dots, \gamma_t$$

are not zero, and $t$ is a non-negative integer. These poles $z = \gamma$ of $W$ occur at zeros of $w$ , and hence are simple since $W = \frac{w'}{w}$ . Furthermore, their residues are

1 . For by its differential equation, $w$ can only have a simple zero at $z = \gamma$ , because otherwise both $w$ and $w'$ would vanish at this regular point of the differential equation, which is impossible.

Therefore, in the neighbourhood of each of the points (42), $W$ can be written as a series

$$(43): \qquad W = \frac{1}{z-\gamma} + \sum_{k=0}^{\infty} C_k(z-\gamma)^k .$$

From the developments (40), (41), and (43), it follows finally that $W$ has one of the explicit forms

$$W = \mp i - \frac{2\lambda}{z} + \sum_{u=1}^{t} \frac{1}{z-\gamma_u} \quad \text{or} \quad W = \mp i - \frac{2\mu}{z} + \sum_{u=1}^{t} \frac{1}{z-\gamma_u} ,$$

and hence, in the neighbourhood of the point at infinity, allows a development of one of the two forms

$$W = \mp i + \frac{-2\lambda+t}{z} + \sum_{k=2}^{\infty} D_k z^{-k} \quad \text{or} \quad W = \mp i + \frac{-2\mu+t}{z} + \sum_{k=2}^{\infty} D_k z^{-k} .$$

Therefore, by (40), either $-2\lambda + t$ or $-2\mu + t$ is equal to $\frac{-2\lambda-2\mu-1}{2}$ , and hence

$2(\lambda-\mu)$ is an odd integer.

Conversely, if this condition is not satisfied, then the Riccati equation (38) has no algebraic solution.

112. We now combine the results proved in the last two sections, but assume that

(44): $\lambda \neq -1, -2, -3, \ldots$ ; $\mu \neq -1, -2, -3, \ldots$ ;
and $2(\lambda-\mu)$ is not an odd integer.

Under this hypothesis, we can assert the following result.

(45): *The differential equation* (32) *for* $K(z;\lambda,\mu)$ *has the algebraic solution* $w \equiv 0$ *exactly when* $\lambda = 0$ *or* $\mu = 0$ . *If* $\lambda$ *is a natural number and* $\binom{\mu}{\lambda} \neq 0$ , *it has the algebraic solution* $w = \sum_{k=1}^{\lambda} A_{2k} z^{-2k}$ , *where the* $A_{2k}$ *are given by* (37) *for* $k = 1, 2, \ldots, \lambda$ ; *similarly, if* $\mu$ *is a natural number and* $\binom{\lambda}{\mu} \neq 0$ , *it has the algebraic solution* $w = \sum_{k=1}^{\mu} A_{2k} z^{-2k}$ , *where the* $A_{2k}$ *are given by* (37) *for* $k = 1, 2, \ldots, \mu$ . *It has otherwise no algebraic solutions. There is furthermore no transcendental integral of the differential equation* (32) *which satisfies an algebraic differential equation of the first order.*

From its definition, the particular integral $K(z;\lambda,\mu)$ of (32) is an entire transcendental function. The result (45) implies therefore that *if the rational numbers* $\lambda$ *and* $\mu$ *satisfy the conditions* (44), *then the functions*

$K(z;\lambda,\mu)$ *and* $K'(z;\lambda,\mu)$ *are algebraically independent over* $C(z)$ .

Hence, by Shidlovski's Theorem 10, we deduce under the same hypothesis that

(46): $K(\alpha;\lambda,\mu)$ *and* $K'(\alpha;\lambda,\mu)$ *are algebraically independent over* $Q$ if $\alpha \neq 0$ *is any algebraic number. Hence both function values are transcendental.*

Of particular interest is the case $\lambda = \nu$ , $\mu = 0$ of this result. It becomes then Siegel's classical result of 1929.

**(47):** *If* $\nu$ *is a rational number distinct from* $-1, -2, -3, \ldots$ *and such that* $2\nu$ *is not an odd integer, and if further* $\alpha \neq 0$ *is an algebraic number, then the two function values*

$$J_\nu(\alpha)\Gamma(\nu+1) \quad \textit{and} \quad J'_\nu(\alpha)\Gamma(\nu+1)$$

*are algebraically independent over* $Q$ *and in particular are transcendental.*

For these two functions can be written in the form

$$J_\nu(z)\Gamma(\nu+1) = (z/2)^\nu K(z;\nu,0) ,$$

$$J'_\nu(z)\Gamma(\nu+1) = \frac{\nu}{2}(z/2)^{\nu-1}K(z;\nu,0) + (z/2)^\nu K'(z;\nu,0) ,$$

and when $\alpha$ is an algebraic number, so are $(\alpha/2)^\nu$ and $\frac{\nu}{2}(\alpha/2)^{\nu-1}$ .

If, in particular, $\nu = n$ is a non-negative integer, $\Gamma(\nu+1) = n!$ also is a rational integer. Therefore the function values

$$J_n(\alpha) \quad \text{and} \quad J'_n(\alpha)$$

themselves are algebraically independent over $Q$ , and each one is transcendental. When the rational number $\nu$ is not an integer, no such assertion about $J_\nu(\alpha)$ and $J'_\nu(\alpha)$ can at present be made because it is then not in general known whether $\Gamma(\nu+1)$ is algebraic or transcendental.

**113.** Siegel (1929) in fact proved a more general result than (47).

Let us change the notation slightly and write

$$K_0(z;\nu) = K(z;\nu,0) .$$

Assume that $\nu$ is not an integer. It is easily verified that the homogeneous linear differential equation

(48): $$w'' + \frac{2\nu+1}{z} w' + w = 0$$

has the two solutions

$$U = K_0(z;\nu) \quad \text{and} \quad V = z^{-2\nu} K_0(z;-\nu) \ .$$

Their Wronski determinant is

$$UV' - U'V = C.\exp\left(- \int \frac{2\nu+1}{z} dz\right) = C.z^{-2\nu-1}$$

where $C$ is a constant which can be determined by substituting for $U$, $V$, $U'$, $V'$ the first terms of their respective series. In this way, it follows that

(49): $$UV' - U'V = -2\nu z^{-2\nu-1} \ .$$

Thus $U$ and $V$ are certainly independent solutions of (48).

The function $-2\nu z^{-2\nu-1}$ on the right-hand side of (49) is algebraic for rational $\nu$. Therefore, *if* $\nu$ *is a rational number not an integer, the four functions*

$$U, V, U' , \text{ and } V'$$

*are algebraically dependent over* $\mathbb{C}(z)$ .

In the opposite direction, we have the following result (Siegel 1929).

**(50):** *If* $\nu$ *is a rational number and* $2\nu$ *is not an integer, then the three functions*

$$U, V, \text{ and } U',$$

*and hence also the three functions*

$$K_0(z;\nu), K_0'(z;\nu), \text{ and } K_0(z;-\nu),$$

*are algebraically independent over* $\mathbb{C}(z)$ .

**114.** Siegel's proof of (50) runs as follows. Denote by $u$ and $v$ an arbitrary pair of independent solutions of (48); then

(51): $$uv' - u'v = cz^{-2\nu-1} ,$$

where $c$ is a constant distinct from zero. Our aim is to prove the more general assertion that also the three functions

$$u, v, \text{ and } u'$$

are algebraically independent over $\mathbb{C}(z)$ . The proof is indirect, and *it will thus be assumed that* $u$, $v$, *and* $u'$ *are algebraically dependent over* $\mathbb{C}(z)$ .

Denote by $L$ and $M$ the two extension fields

$$L = \mathbf{C}(z, z^{-2\nu-1}) \quad \text{and} \quad M = L(u, u')$$

of $\mathbf{C}(z)$ . By (45), $u$ and $u'$ are algebraically independent over $\mathbf{C}(z)$ and hence also over $L$ . On the other hand, it follows from the above assumption that $v$ is algebraic over $M$ and that therefore the same is true for the quotient

$$s = v/u \ .$$

Hence there exists an irreducible polynomial

$$P(w) = w^n + P_1 w^{n-1} + \ldots + P_n$$

of positive degree $n$ in $w$ and with coefficients in $M$ such that

(52): $$P(s) = 0 \ .$$

By the relation (51), $s'$ has the form

(53): $$s' = X(z)/u^2$$

where $X(z)$ denotes the function

$$X(z) = cz^{-2\nu-1}$$

which lies in $L$ . The most general integral of the differential equation (53) for $s$ is given by $s + \gamma$ where $\gamma$ is an arbitrary constant.

Put

$$\Pi(w) = X(z)u^{-2}\frac{\partial P(w)}{\partial w} + \sum_{k=1}^{n}\frac{dP_k}{dz}w^{n-k} \ ;$$

this is a polynomial of degree at most $n - 1$ in $w$ , with coefficients in $M$ . From its definition,

(54): $$\Pi(s+\gamma) = \frac{dP(s+\gamma)}{dz} \quad \text{for every } \gamma \text{ in } \mathbf{C} \ ,$$

hence, for $\gamma = 0$ , by (52), $\Pi(s) = 0$ .

This equation for $s$ is of lower degree than the irreducible equation (52); it can then hold only if $\Pi(w)$ vanishes identically in $w$ . Thus it follows from (54) that identically in $z$ and $\gamma$ ,

(55): $$P(s+\gamma) = g(\gamma)$$

where $g(\gamma)$ is a certain function of $\gamma$ which is independent of $z$ . Therefore, from the explicit form of $P(w)$ , $g(\gamma)$ is a polynomial

$$g(\gamma) = \gamma^n + g_1\gamma^{n-1} + \ldots + g_n$$

in $\gamma$ with coefficients $g_1, \ldots, g_n$ in $C$. On comparing the coefficients of $\gamma^{n-1}$ on both sides of the identity (55), we find that

$$s = (g_1 - P_1)/n$$

and so establish that $n = 1$ and that $s$ lies in the field $M$.

115. The quotient $s = v/u$ can therefore be written in the form

$$s = N/D ,$$

where $N = N(u,u')$ and $D = D(u,u')$ are polynomials in $L[u,u']$. Here, since $s \not\equiv 0$, neither $N$ nor $D$ is identically zero. Denote by $n$ and $d$ the total degrees of $N$ and $D$, respectively, and write these polynomials as the sums

$$N = \sum_{i=0}^{n} N_i \quad \text{and} \quad D = \sum_{j=0}^{d} D_j$$

of their homogeneous parts $N_i$ and $D_j$ of dimensions $i = 0, 1, \ldots, n$ and $j = 0, 1, \ldots, d$, respectively. Some of these homogeneous parts may vanish identically, but $N_n$ and $D_d$ are not identically zero.

The function $u$ satisfies the differential equation

$$u'' + \frac{2\nu+1}{z} u' + u = 0 .$$

Hence

$$\frac{dN_i}{dz} = \frac{\partial N_i}{\partial z} + \frac{\partial N_i}{\partial u} u' - \frac{\partial N_i}{\partial u'} \left(\frac{2\nu+1}{z} u' + u\right) , \quad = N_i^* \quad \text{say,}$$

$$\frac{dD_j}{dz} = \frac{\partial D_j}{\partial z} + \frac{\partial D_j}{\partial u} u' - \frac{\partial D_j}{\partial u'} \left(\frac{2\nu+1}{z} u' + u\right) , \quad = D_j^* \quad \text{say,}$$

and therefore also

$$(56): \quad \frac{dN}{dz} = \sum_{i=0}^{n} N_i^* \quad \text{and} \quad \frac{dD}{dz} = \sum_{j=0}^{d} D_j^* .$$

It is evident from the definitions of $N_i^*$ and $D_j^*$ that these expressions are homogeneous polynomials in $u$ and $u'$ which either vanish identically, or are of the same dimensions $i$ as $N_i$, or $j$ as $D_j$, respectively. The formulae (56) thus contain the decompositions of the derivatives of $N$ and $D$ into their homogeneous parts, and the total degrees of $\frac{dN}{dz}$ and of $\frac{dD}{dz}$ cannot exceed $n$ and $d$, respectively.

Next we find that

$$s' = \frac{d(N/D)}{dz} = \left(D\frac{dN}{dz} - \frac{dD}{dz}N\right)D^{-2} = \left[\sum_{i=0}^{n}\sum_{j=0}^{d}(N_i^* D_j - N_i D_j^*)\right]\left[\sum_{j=0}^{d}\sum_{k=0}^{d} D_j D_k\right]^{-1},$$

so that by (53),

(57): $$u^2 \sum_{i=0}^{n}\sum_{j=0}^{d}(N_i^* D_j - N_i D_j^*) = X(z)\sum_{j=0}^{d}\sum_{k=0}^{d} D_j D_k .$$

Here the total degree of the right-hand side is equal to $2d$, while that of the left-hand side evidently does not exceed $2 + n + d$. Therefore $2+n+d \geq 2d$ and hence

$$n \geq d-2 .$$

We now distinguish two cases. Firstly let $n = d - 2$. Then, on comparing the terms of highest dimension on the two sides of (57), it follows that

(58): $$u^2(N_n^* D_d - N_n D_d^*) = X(z)D_d^2 .$$

This means that the rational function

$$t = t(u,u') = \frac{N_n(u,u')}{D_d(u,u')}$$

of $u$ and $u'$ satisfies the differential equation

(59): $$t' = X(z)u^{-2}$$

which is the same as the equation (53) for $s$.

If the differentiation with respect to $z$ is carried out in all terms of (58), and if afterwards the second derivative $u''$ is replaced by its value

$$u'' = -\frac{2\nu+1}{z}u' - u ,$$

a first order algebraic differential equation for $u$ with coefficients in $L$ is obtained. By (45), $u$ and $u'$ are, however, algebraically independent over $L$. This first order differential equation reduces therefore to an identity, and it follows that the relations (58) and (59) remain valid if the function $u$ is replaced by an arbitrary solution $w \not\equiv 0$ of

$$w'' + \frac{2\nu+1}{z}w' + w = 0 .$$

By hypothesis, $u$ and $v$ form an independent pair of solutions of this differential equation; hence $w$ has the form

$$w = \alpha u + \beta v ,$$

where $\alpha$ and $\beta$ are two arbitrary constants not both zero. For every such pair of constants $\alpha$, $\beta$ the quotient

(60): $$T(z;\alpha,\beta) = \frac{N_n(\alpha u+\beta v,\alpha u'+\beta v')}{D_d(\alpha u+\beta v,\alpha u'+\beta v')}$$

satisfies the differential equation (59) and hence differs from $s$ , a solution of (53), only by a quantity independent of $z$ . Thus

(61): $$T(z;\alpha,\beta) = s + \tau(\alpha,\beta) ,$$

where $\tau(\alpha,\beta)$ is a certain function of $\alpha$ and $\beta$ which does not depend on $z$ . Since

$$s = v/u ,$$

$\tau(\alpha,\beta)$ has by (60) and (61) the explicit form

(62): $$\tau(\alpha,\beta) = \frac{uN_n(\alpha u+\beta v,\alpha u'+\beta v')-vD_d(\alpha u+\beta v,\alpha u'+\beta v')}{uD_d(\alpha u+\beta v,\alpha u'+\beta v')} .$$

It follows that $\tau(\alpha,\beta)$ is a rational function of $\alpha$ and $\beta$ with coefficients in $M$ . In fact, since $\tau(\alpha,\beta)$ is independent of $z$ , its coefficients are constants in $\mathrm{C}$ .

Both the numerator and the denominator of the right-hand side of (62) are polynomials in $\alpha$ and $\beta$ of the exact total degree $d$ , because $n = d - 2$ . The denominator is homogeneous, but the numerator is not. Hence $\tau(\alpha,\beta)$ certainly is not a constant, but depends in a non-trivial way on $\alpha$ and $\beta$ .

Denote by $\delta(\alpha,\beta)$ the greatest common divisor of the two homogeneous polynomials $N_n(\alpha u+\beta v,\alpha u'+\beta v')$ and $D_d(\alpha u+\beta v,\alpha u'+\beta v')$ in $\alpha$ and $\beta$ , and put

$$\mathrm{N}(\alpha u+\beta v,\alpha u'+\beta v') = N_n(\alpha u+\beta v,\alpha u'+\beta v')\delta(\alpha,\beta)^{-1}$$

and

$$\mathrm{D}(\alpha u+\beta v,\alpha u'+\beta v') = D_d(\alpha u+\beta v,\alpha u'+\beta v')\delta(\alpha,\beta)^{-1} ,$$

so that

(63): $$\tau(\alpha,\beta) = \frac{u\mathrm{N}(\alpha u+\beta v,\alpha u'+\beta v')-v\mathrm{D}(\alpha u+\beta v,\alpha u'+\beta v')}{u\mathrm{D}(\alpha u+\beta v,\alpha u'+\beta v')} .$$

Here $\mathrm{D}(\alpha u+\beta v,\alpha u'+\beta v')$ is a homogeneous polynomial in $\alpha$ and $\beta$ of dimension not less than 2 , and it is relatively prime to $\mathrm{N}(\alpha u+\beta v,\alpha u'+\beta v')$ . There exist then two constants $\alpha = A$ and $\beta = B$ in $\mathrm{C}$ , not both zero, such that

$$\mathrm{D}(Au+Bv,Au'+Bv') \text{ , but not } \mathrm{N}(Au+Bv,Au'+Bv') ,$$

vanishes identically as function of $z$ . Now, by (63),

(64): $$u\mathrm{D}(Au+Bv,Au'+Bv')\tau(A,B) = u\mathrm{N}(Au+Bv,Au'+Bv') - v\mathrm{D}(Au+Bv,Au'+Bv') ,$$

so that

$$N(Au+Bv, Au'+Bv') = 0 .$$

This, however, is a first order algebraic differential equation for $w = Au + Bv \not\equiv 0$ which is not an identity, contrary to what was proved in (45).

Since, then, the case $n = d - 2$ is impossible, necessarily $n \geq d-1$. The identity (57) therefore now implies that

$$N_n^* D_d - N_n D_d^* = 0 ,$$

and hence that

$$\frac{d}{dz}(N_n/D_d) = 0 .$$

Therefore $N_n/D_d$ is a constant, say $c$, and hence $n = d$. The new function

$$s_1 = s - c = N^0/D , \text{ where } N^0 = N - cD ,$$

again has the property

$$s_1' = X(z)u^{-2}$$

and is a rational function of $u$ and $v$. But the total degree of its numerator, $n^0$ say, is now less than $n = d$ and, by what has just been said, cannot have the value $d - 1$; hence necessarily $n^0 = d - 2$. We therefore come back to the case which had already been considered and found to be impossible.

Thus it has been proved that if $\nu$ is rational, and $2\nu$ is not an integer, then the three functions $u$, $v$, and $u'$ are algebraically independent.

For almost trivial reasons, an analogous result holds when $\nu$ is a non-negative integer. For let us take for $u$ the integral $u = K_0(z;\nu)$; then $u$ and $u'$ are regular functions at the origin $z = 0$ and are algebraically independent over $\mathbb{C}(z)$. It can now be proved that every independent integral $v$ must have a logarithmic singularity at $z = 0$, hence certainly is algebraically independent of $u$ and $u'$ over $\mathbb{C}(z)$. This second integral naturally is no longer now a Siegel $E$-function.

**116.** Let us finally combine Theorem 10 of Shidlovski with the result (50). We obtain then the following theorem, which was first obtained by Siegel (1929).

**(65):** *Let $\nu$ be a rational number such that $2\nu$ is not an integer, and let $\alpha \neq 0$ be any algebraic number. Then the three function values*

$$K_0(\alpha;\nu),\ K_0'(\alpha;\nu),\ \text{and}\ K_0(\alpha;-\nu)$$

*are algebraically independent over $\mathbb{Q}$.*

This property implies that also the three products

$$J_\nu(\alpha)\Gamma(\nu+1),\ J'_\nu(\alpha)\Gamma(\nu+1)\ ,\ \text{and}\ \ J_{-\nu}(\alpha)\Gamma(-\nu+1)$$

are algebraically independent over $Q$. Further, since $\nu$ is a rational number,

$$\Gamma(\nu+1)\Gamma(-\nu+1) = \frac{\nu\pi}{\sin \nu\pi}$$

is an algebraic multiple of $\pi$, so that we obtain the rather curious result that the two products

$$\pi J_\nu(\alpha)J_{-\nu}(\alpha) \quad \text{and} \quad J'_\nu(\alpha)J_{-\nu}(\alpha)$$

are likewise algebraically independent over $Q$. This property can be extended to the case when $2\nu$ is an odd integer.

117. The Siegel $E$-function

$$K_0(z;\nu) = \sum_{n=0}^{\infty} \frac{(-z^2/4)^n}{n![\nu+1,n]}$$

is a solution of the differential equation

$$w'' + \frac{2\nu+1}{z} w' + w = 0 \ .$$

Let us now differentiate $K_0(z;\nu)$ repeatedly with respect to the parameter $\nu$ and after this differentiation put $\nu = 0$. We obtain then the new functions

(66): $$K_k(z) = \frac{1}{k!} \left(\frac{\partial}{\partial\nu}\right)^k K_0(z;\nu)\Big|_{\nu=0} \qquad (k = 0,1,2,\ldots)$$

which evidently form a solution of the following infinite system of homogeneous linear differential equations,

(67): $$w''_0 + \frac{1}{z} w'_0 + w_0 = 0 \ , \quad w''_k + \frac{1}{z} w'_k + w_k + \frac{2}{z} w'_{k-1} = 0 \qquad (k = 1,2,3,\ldots) \ .$$

On carrying out the differentiations in (66), we obtain for $K_k(z)$ the explicit formula

(68): $$K_k(z) = \sum_{n=0}^{\infty} p_k(0;n) \frac{(-z^2/4)^n}{k!(n!)^2} \ ,$$

where $p_k(\beta;n)$ denotes the expression which was defined in §104. A proof similar to that in §104 leads to the result that *all the functions* $K_k(z)$ *are Siegel E-functions.*

We arrive then at the non-trivial problem of deciding which of the functions

(69): $$K_k(z),\ K'_k(z) \qquad (k = 0,1,2,\ldots)$$

are algebraically independent over $\mathbb{C}(z)$. One can show that these infinitely many functions are, in fact, connected by infinitely many independent quadratic identities with coefficients in $\mathbb{C}(z)$. The simplest of these identities are

$$z(K_0K_1'-K_0'K_1) + K_0^2 - 1 = 0$$

and

$$z(K_0K_3'-K_1K_2'+K_2K_1'-K_3K_0') + \left(2K_0K_2-K_1^2\right) = 0 ,$$

of which the first is due to Belogrivov (1967) and the second one to myself (Mahler 1968). See also Väänänen (1972, 1973) for more general results.

It can be proved (Mahler *loc. cit.*) that, for example, the six functions

(70): $$K_0(z),\ K_0'(z),\ K_1(z),\ K_2(z),\ K_2'(z),\ K_3(z)$$

are algebraically independent over $\mathbb{C}(z)$, while naturally $K_1'(z)$ and $K_3'(z)$ are algebraically dependent over $\mathbb{C}(z)$ on these six functions. Theorem 10 therefore easily implies the following result.

(71): *If* $\alpha \neq 0$ *is any algebraic number, then the six function values*

$$K_0(\alpha),\ K_0'(\alpha),\ K_1(\alpha),\ K_2(\alpha),\ K_2'(\alpha),\ K_3(\alpha)$$

*are algebraically independent over* $\mathbb{Q}$.

This result can be put in an equivalent form which is perhaps more interesting. For this purpose, put

$$C_k(z) = \frac{1}{k!}\left(\frac{\partial}{\partial\nu}\right)^k J_\nu(z)\Big|_{\nu=0} \qquad (k = 0,1,2,\ldots) ,$$

so that, in particular,

$$C_0(z) = J_0(z) , \quad C_1(z) = \frac{2}{\pi} Y_0(z) ,$$

where $Y_0(z)$ is the Bessel function of the second kind. Further, let

$$c_1 = \log(\alpha/2) + \gamma , \quad c_2 = \left(c_1^2+\zeta(2)\right)/2 , \quad c_3 = \left(c_1^3+3c_1\zeta(2)+2\zeta(3)\right)/6 ,$$

where $\alpha \neq 0$ is again an arbitrary algebraic number, $\gamma$ denotes Euler's constant, and $\zeta(s)$ is the Riemann Zeta function. With this notation, put

$$A_0 = C_0(\alpha) , \quad A_1 = C_1(\alpha) - c_1C_0(\alpha) ,$$

$$A_2 = C_2(\alpha) - c_1C_1(\alpha) + c_2C_0(\alpha) , \quad A_3 = C_3(\alpha) - c_1C_2(\alpha) + c_2C_1(\alpha) - c_3C_0(\alpha) ,$$

and denote by $A_0', A_1', A_2'$, and $A_3'$ the expressions analogous to $A_0, A_1, A_2$, and $A_3$ which are obtained when $C_0(\alpha), C_1(\alpha), C_2(\alpha)$, and $C_3(\alpha)$ are replaced by

$C_0'(\alpha)$, $C_1'(\alpha)$, $C_2'(\alpha)$ , and $C_3'(\alpha)$ , respectively.

From (71), we easily deduce the following theorem.

(72): *The six numbers*

$$A_0, A_0', A_1, A_2, A_2', A_3$$

*are algebraically independent over* $\mathbb{Q}$ , *and so are any other six of the eight numbers*

$$A_0, A_0', A_1, A_1', A_2, A_2', A_3 \text{ , and } A_3' .$$

We thus find expressions involving Euler's constant and the constants $\zeta(2) = \pi^2/6$ and $\zeta(3)$ of which the transcendency can be proved, and algebraic operations allow one to construct further numbers of this kind. By way of example,

$$\frac{\pi}{2}\left(Y_0(\alpha)/J_0(\alpha)\right) - \gamma$$

is a transcendental number.

For a detailed proof and for infinitely many analogous results, I refer to my paper Mahler (1968) and to Väänänen (1972).

The examples of this chapter show how powerful the general theory of Shidlovski is. He himself and a number of his students have given many other applications of his general theorems, and the reader is referred to their work.

# CHAPTER 8

## FORMAL POWER SERIES AS SOLUTIONS OF ALGEBRAIC DIFFERENTIAL EQUATIONS

**118.** This chapter will deal with a theorem by J. Popken (1935) on the coefficients of formal power series which satisfy algebraic differential equations, a theorem which allows interesting applications to the transcendency of certain function values from the theory of elliptic functions.

Popken's theorem is not as well known as it deserves. It is therefore appropriate to include it in these notes.

**119.** For the present, let $K$ be an arbitrary field of characteristic $0$. With a slight change of the previous notation, denote by $K^*$ the ring of all formal power series

$$f = \sum_{h=0}^{\infty} f_h z^h \,, \quad g = \sum_{h=0}^{\infty} g_h z^h, \ \ldots$$

and so on, with coefficients $f_h, g_h, \ldots$ in $K$. Here sum and product are defined by

$$f + g = \sum_{h=0}^{\infty} (f_h + g_h) z^h \,, \quad fg = \sum_{h=0}^{\infty} \left( \sum_{k=0}^{h} f_k g_{h-k} \right) z^h$$

The elements $a$ of $K$ are identified with the power series

$$a = a + \sum_{h=1}^{\infty} 0 . z^h$$

and play the role of constants.

Differentiation in $K^*$ is defined formally by

$$\frac{d^k f}{dz^k} = f^{(k)} = \sum_{h=k}^{\infty} h(h-1) \ldots (h-k+1) f_h z^{h-k} \,,$$

a notation still used for $k = 0$ when

$$f^{(0)} = f \,.$$

In particular,

$$\frac{df}{dz} = 0 \text{ if and only if } f \equiv a \in K \,.$$

The usual rules for the derivatives of sum, difference, and product remain valid for the elements of $K^*$ .

An important mapping from $K^*$ into $K$ is defined by the formal substitution $z = 0$ for which we use the notations

$$f \to f(0) = f\Big|_{z=0} = f_0 .$$

More generally, in terms of the coefficients of $f$ ,

$$f^{(k)}(0) = f^{(k)}\Big|_{z=0} = k!f_k .$$

**120.** Our aim is to study formal power series

$$f = \sum_{h=0}^{\infty} f_h z^h$$

in $K^*$ which satisfy algebraic differential equations

(1): $$F\big((f)\big) \equiv F(z;f,f',\ldots,f^{(m)}) = 0 .$$

Here $F(z;w_0,w_1,\ldots,w_m) \not\equiv 0$ is a polynomial in indeterminates $z$, $w_0$, $w_1$, ..., $w_m$ with coefficients in some extension field of $K$ . A method already applied several times in previous chapters shows immediately that $f$ then satisfies also a differential equation of type (1) over $K$ itself, that is, with coefficients in $K$ . Hence only this case need be considered from now on.

We are further allowed to assume that (1) is an algebraic differential equation for $f$ over $K$ which has lowest possible order $m$ and which, in addition, is of lowest degree in $f^{(m)}$ among all algebraic differential equations over $K$ of this order $m$ . Since factors of $F$ not involving $f^{(m)}$ may be divided out, without loss of generality the polynomial $F(z;w_0,w_1,\ldots,w_m)$ may be assumed to be irreducible over $K$ .

The degree of (1) in $f^{(m)}$ is necessarily positive. The order $m$ may be zero, and then (1) becomes an algebraic equation for $f$ ; (1) is a proper differential equation for $f$ only if $m \geq 1$ .

It is convenient to use the notations

$$F_\mu(z;w_0,w_1,\ldots,w_m) = \frac{\partial}{\partial w_\mu} F(z;w_0,w_1,\ldots,w_m) \qquad (\mu = 0,1,\ldots,m)$$

and

$$F_{(\mu)}\big((w)\big) = F_\mu(z;w,w',\ldots,w^{(m)}) \qquad (\mu = 0,1,\ldots,m)$$

where $w$ may be any series in $K^*$. The assumption about $F$ imply then that

(2): $$F_{(m)}((f)) \equiv F_m(z;f,f',\ldots,f^{(m)}) \neq 0 .$$

For $F((w))$ depends explicitly on $w^{(m)}$; therefore

$$F_{(m)}((f)) = 0$$

would be a differential equation for $f$ of either lower order, or of the same order but of lower degree, than the differential equation (1), contrary to the hypothesis.

**121.** The differential operator $F((w))$ can be written in the explicit form

(3): $$F((w)) = \sum_{(\kappa)} p_{(\kappa)}(z) w^{(\kappa_1)} \ldots w^{(\kappa_N)} .$$

Here the summation

$$\sum_{(\kappa)}$$

extends over all ordered systems $(\kappa) = (\kappa_1,\ldots,\kappa_N)$ of integers for which

(4): $$0 \leq \kappa_1 \leq m, \ \ldots, \ 0 \leq \kappa_N \leq m; \ \kappa_1 \leq \kappa_2 \leq \ldots \leq \kappa_N; \ 0 \leq N \leq n ,$$

where $n$ is a fixed positive integer, and the coefficients $p_{(\kappa)}(z)$ are polynomials in $K[z]$. The integer $N$ is variable with $(\kappa)$, and there is exactly one *improper system* with $N = 0$ which will be denoted by $(\omega)$. The corresponding term

$$p_{(\omega)}(z)$$

in (3) has no factors $w^{(j)}$ and is a polynomial in $z$ alone.

**122.** We require an explicit expression for the derivatives

(5): $$F^{(h)}((w)) = \left(\frac{d}{dz}\right)^h F((w)) \qquad (h = 1,2,3,\ldots)$$

of $F((w))$ and shall obtain it by means of the following lemma.

*Let $h \geq 1$ and $N \geq 0$ be any two integers, and let $w_0, w_1, \ldots, w_N$ be $N + 1$ arbitrary series in $K^*$. Then*

(6): $$\left(\frac{d}{dz}\right)^h (w_0 w_1 \ldots w_N) = h! \sum_{\lambda_0,\lambda_1,\ldots,\lambda_N} \frac{w_0^{(\lambda_0)}}{\lambda_0!} \frac{w_1^{(\lambda_1)}}{\lambda_1!} \cdots \frac{w_N^{(\lambda_N)}}{\lambda_N!}$$

*where the summation extends over all ordered systems of integers $\lambda_0, \lambda_1, \ldots, \lambda_N$ such that*

(7): $$\lambda_0 \geq 0,\ \lambda_1 \geq 0,\ \ldots,\ \lambda_N \geq 0;\ \lambda_0 + \lambda_1 + \ldots + \lambda_N = h\ .$$

Proof. The assertion is evidently true if $h = 1$ ; assume it has already been established for some order $h \geq 1$ . We show that it is then valid also for the order $h + 1$ and hence is true for all orders $h$ .

On differentiating (6) once more,

$$\left(\frac{d}{dz}\right)^{h+1}(w_0 w_1 \ldots w_N) = h! \sum_{\lambda_0,\lambda_1,\ldots,\lambda_N} \sum_{\nu=0}^{N} \frac{w_0^{(\lambda_0)}}{\lambda_0!} \cdots \frac{w_\nu^{(\lambda_\nu+1)}}{\lambda_\nu!} \cdots \frac{w_N^{(\lambda_N)}}{\lambda_N!} =$$

$$= h! \sum_{\mu_0,\mu_1,\ldots,\mu_N} \left(\sum_{\nu=0}^{N} \mu_\nu\right) \frac{w_0^{(\mu_0)}}{\mu_0!} \frac{w_1^{(\mu_1)}}{\mu_1!} \cdots \frac{w_N^{(\mu_N)}}{\mu_N!}\ ,$$

where the new summation extends over all ordered systems of integers $\mu_0, \mu_1, \ldots, \mu_N$ for which

$$\mu_0 \geq 0,\ \mu_1 \geq 0,\ \ldots,\ \mu_N \geq 0;\ \mu_0 + \mu_1 + \ldots + \mu_N = h + 1\ .$$

Now

$$h! \sum_{\nu=0}^{N} \mu_\nu = (h+1)!\ ,$$

whence the assertion.

123. The lemma just proved will now be applied to all the separate terms

(8): $$p_{(\kappa)}(z) w^{(\kappa_1)} \ldots w^{(\kappa_N)}$$

in the representation (3) of $F\big((w)\big)$ . It follows immediately from the lemma that

(9): $$F^{(h)}\big((w)\big) = h! \sum_{(\kappa)} \sum_{[\lambda]} \frac{p_{(\kappa)}^{(\lambda_0)}(z)}{\lambda_0!} \frac{w^{(\kappa_1+\lambda_1)}}{\lambda_1!} \cdots \frac{w^{(\kappa_N+\lambda_N)}}{\lambda_N!}\ .$$

Here the inner sum

$$\sum_{[\lambda]}$$

is extended over all ordered systems $[\lambda] = [\lambda_0, \lambda_1, \ldots, \lambda_N]$ of integers which satisfy the conditions

(7): $$\lambda_0 \geq 0,\ \lambda_1 \geq 0,\ \ldots,\ \lambda_N \geq 0;\ \lambda_0 + \lambda_1 + \ldots + \lambda_N = h\ ,$$

while $N$ has the same value as in the term (8) from which this sum arises. There is in particular exactly one term

$$p_{(\omega)}^{(h)}(z)$$

on the right-hand side of (9) for which $N = 0$. This term is independent of $w$ and vanishes as soon as $h$ exceeds the degree of the polynomial $p_{(\omega)}(z)$.

**124.** Denote next by $j$ any integer in the interval

(10): $$0 \le j \le \left[\frac{h-1}{2}\right],$$

put

(11): $$k = h + m - j,$$

and denote by

(12): $$F^{(h,k)}((w))w^{(k)}$$

the sum of all terms on the right-hand side of (9) which have at least one factor $w^{(k)}$. Further, let

(13): $$F_{(\kappa)}^{(h,k)}((w))w^{(k)}$$

be the sum of all those contributions to $F^{(h,k)}((w)).w^{(k)}$ which come from the $h$-th derivative

(14): $$\left(\frac{d}{dz}\right)^h\left[p_{(\kappa)}(z)w^{(\kappa_1)}\dots w^{(\kappa_N)}\right] = h! \sum_{[\lambda]} \frac{p_{(\kappa)}^{(\lambda_0)}(z)}{\lambda_0!}\frac{w^{(\kappa_1+\lambda_1)}}{\lambda_1!}\dots\frac{w^{(\kappa_N+\lambda_N)}}{\lambda_N!}$$

of the term

$$t_{(\kappa)} = p_{(\kappa)}(z)w^{(\kappa_1)}\dots w^{(\kappa_N)}$$

in the representation (3) of $F((w))$. Evidently

(15): $$F^{(h,k)}((w)) = \sum_{(\kappa)} F_{(\kappa)}^{(h,k)}((w)),$$

where trivially

(16): $$F_{(\omega)}^{(h,k)}((w)) = 0.$$

It suffices therefore to consider only those terms $t_{(\kappa)}$ for which $(\kappa) \ne (\omega)$ and hence $1 \le N \le n$.

**125.** Denote by $\nu$ any suffix $1, 2, \dots, N$, and by $\nu'$ any such suffix which is distinct from $\nu$ . It is trivial that the binomial coefficient

(17): $$\binom{h}{k-\kappa_\nu} = 0 \text{ if } k - \kappa_\nu < 0 \text{ or } k - \kappa_\nu > h .$$

Next let $\nu$ be such that

(18): $$0 \le k-\kappa_\nu \le h .$$

There exist then systems $[\lambda] = [\lambda_0, \lambda_1, \ldots, \lambda_N]$ of $N + 1$ integers with the properties

(7): $$\lambda_0 \ge 0,\ \lambda_1 \ge 0,\ \ldots,\ \lambda_N \ge 0;\ \lambda_0 + \lambda_1 + \ldots + \lambda_N = h ,$$

and

(19): $$\kappa_\nu + \lambda_\nu = k .$$

Since then

$$\lambda_\nu = k - \kappa_\nu = (h-j) + (m-\kappa_\nu) \ge h - j > \frac{h}{2} ,$$

and therefore

$$\lambda_{\nu'} < \frac{h}{2},\quad \kappa_{\nu'} + \lambda_{\nu'} < \frac{h}{2} + m = h + m - \frac{h}{2} \le h + m - j = k ,$$

the corresponding term

$$T_{(\kappa),[\lambda]} = h! \, \frac{P_{(\kappa)}^{(\lambda_0)}(z)}{\lambda_0!} \, \frac{w^{(\kappa_1+\lambda_1)}}{\lambda_1!} \cdots \frac{w^{(\kappa_N+\lambda_N)}}{\lambda_N!}$$

in (14) has exactly *one* factor $w^{(k)}$ and no more. Hence the contribution to $F_{(\kappa)}^{(h,k)}\big((w)\big)$ of this term $T_{(\kappa),[\lambda]}$ for the suffix $\nu$ and the system $[\lambda]$ is equal to

$$\frac{\partial T_{(\kappa),[\lambda]}}{\partial w^{(k)}} = \frac{h!}{\lambda_\nu!} \, \frac{P_{(\kappa)}^{(\lambda_0)}(z)}{\lambda_0!} \prod_{\nu'} \frac{w^{(\kappa_{\nu'}+\lambda_{\nu'})}}{\kappa_{\nu'}!} .$$

Next, by Lemma 1,

$$\left(\frac{d}{dz}\right)^{h-k+\kappa_\nu}\left[P_{(\kappa)}(z) \prod_{\nu'} w^{(\kappa_{\nu'})}\right] = (h-k+\kappa_\nu)! \sum_{[\lambda]}{}' \frac{P_{(\kappa)}^{(\lambda_0)}(z)}{\lambda_0!} \prod_{\nu'} \frac{w^{(\kappa_{\nu'}+\lambda_{\nu'})}}{\lambda_{\nu'}!} ,$$

where the summation $\sum_{[\lambda]}'$ extends only over those systems $[\lambda]$ for which both conditions (7) and (19) are satisfied. From this equation, it follows that

$$\sum_{[\lambda]}{}' \frac{\partial T_{(\kappa),[\lambda]}}{\partial w^{(k)}} = \binom{h}{k-\kappa_\nu}\left(\frac{d}{dz}\right)^{h-k+\kappa_\nu}\left[p_{(\kappa)}(z)\prod_{\nu'} w^{(\kappa_{\nu'})}\right] =$$

$$= \binom{h}{k-\kappa_\nu}\left(\frac{d}{dz}\right)^{h-k+\kappa_\nu} \frac{\partial}{\partial w^{(\kappa_\nu)}}\left[p_{(\kappa)}(z) w^{(\kappa_1)} \ldots w^{(\kappa_N)}\right] ,$$

whence, on summing over $\nu = 1, 2, \ldots, N$ ,

$$F^{(h,k)}_{(\kappa)}((w)) = \sum_{\nu=1}^{N} \binom{h}{k-\kappa_\nu}\left(\frac{d}{dz}\right)^{h-k+\kappa_\nu} \frac{\partial}{\partial w^{(\kappa_\nu)}}\left[p_{(\kappa)}(z) w^{(\kappa_1)} \ldots w^{(\kappa_N)}\right] ,$$

because, by (17), the terms for which $\nu$ does not satisfy the conditions (18) are equal to zero.

This formula can finally be written as

$$(20): \qquad F^{(h,k)}_{(\kappa)}((w)) = \sum_{\mu=0}^{m} \binom{h}{k-\mu}\left(\frac{d}{dz}\right)^{h-k+\mu} \frac{\partial}{\partial w^{(\mu)}}\left[p_{(\kappa)}(z) w^{(\kappa_1)} \ldots w^{(\kappa_N)}\right] ,$$

since evidently

$$\frac{\partial}{\partial w^{(\mu)}}\left[p_{(\kappa)}(z) w^{(\kappa_1)} \ldots w^{(\kappa_N)}\right] = \sum_{\nu} \frac{\partial}{\partial w^{(\kappa_\nu)}}\left[p_{(\kappa)}(z) w^{(\kappa_1)} \ldots w^{(\kappa_N)}\right] ,$$

where $\nu$ in $\sum_{\nu}$ runs over all suffixes $1, 2, \ldots, N$ for which $\kappa_\nu = \mu$ .

On combining the two relations (15) and (20) with the representation (3) of $F((w))$ and making use of the notation $F_{(\mu)}((w))$ introduced in §120, we arrive at the simple formula

$$(21): \qquad F^{(h,k)}((w)) = \sum_{\mu=0}^{m} \binom{h}{k-\mu}\left(\frac{d}{dz}\right)^{h-k+\mu} F_{(\mu)}((w))$$

which is due to Kakeya (1915).

**126.** The basic identities (9) and (21) hold for all series $w$ in $K^*$ . In particular, let

$$w = f .$$

Since $F((f)) = 0$ , we obtain then firstly the equations

$$(22): \quad F^{(h)}((f)) = h! \sum_{(\kappa)} \sum_{[\lambda]} \frac{p^{(\lambda_0)}_{(\kappa)}(z)}{\lambda_0!} \frac{f^{(\kappa_1+\lambda_1)}}{\lambda_1!} \cdots \frac{f^{(\kappa_N+\lambda_N)}}{\lambda_N!} = 0 \qquad (h = 1,2,3,\ldots) ,$$

and secondly, for all $h = 1, 2, 3, \ldots$ and all $j = 0, 1, \ldots, \left[\frac{h-1}{2}\right]$ , the formula

(23): $$F^{(h,k)}((f)) = \sum_{\mu=0}^{m} \binom{h}{k-\mu} \left(\frac{d}{dz}\right)^{h-k+\mu} F_{(\mu)}((f))$$

for the coefficient of $f^{(k)} = f^{(h+m-j)}$ in (22).

In (22) and (23) put $z = 0$ . Since

$$P_{(\kappa)}^{(\lambda_0)}(z)\Big|_{z=0} = P_{(\kappa)}^{(\lambda_0)}(0) \quad \text{and} \quad f^{(l)}\Big|_{z=0} = l!f_l ,$$

we find that

(24): $$h! \sum_{(\kappa)} \sum_{[\lambda]} \frac{P_{(\kappa)}^{(\lambda_0)}(0)}{\lambda_0!} \frac{(\kappa_1+\lambda_1)!}{\lambda_1!} \cdots \frac{(\kappa_N+\lambda_N)!}{\lambda_N!} f_{\kappa_1+\lambda_1} \cdots f_{\kappa_N+\lambda_N} = 0$$
$$(h = 1,2,3,\ldots) ,$$

and here the coefficients of $k!f_k$ on the left-hand side is given by

(25): $$F^{(h,k)}((f))\Big|_{z=0} = \sum_{\mu=0}^{m} \binom{h}{k-\mu} \left(\frac{d}{dz}\right)^{h-k+\mu} F_{(\mu)}((f))\Big|_{z=0} .$$

The $m+1$ expressions $F_{(\mu)}((f))$ are elements of $K^*$ ; let

(26): $$F_{(\mu)}((f)) = \sum_{h=0}^{\infty} F_{\mu,h} z^h \qquad (\mu = 0,1,\ldots,m)$$

be their formal power series with coefficients in $K$ . Since

(2): $$F_{(m)}((f)) \neq 0 ,$$

not all the coefficients $F_{m,h}$ are zero, and there exists an integer $t \geq 0$ such that

(27): $$F_{m,0} = F_{m,1} = \ldots = F_{m,t-1} = 0 , \text{ but } F_{m,t} \neq 0 .$$

From the power series (26),

$$\left(\frac{d}{dz}\right)^{h-k+\mu} F_{(\mu)}((f))\Big|_{z=0} = (h-k+\mu)!F_{\mu,h-k+\mu} \qquad (\mu = 0,1,\ldots,m) ,$$

whence, by (25),

$$F^{(h,k)}((f))\Big|_{z=0} = \sum_{\mu=0}^{m} \binom{h}{k-\mu} (h-k+\mu)!F_{\mu,h-k+\mu} .$$

Replace here $\mu$ by $m-\mu$ ; since $k = h+m-j$ , the formula then takes the simpler form

(28): $$F^{(h,k)}\big((f)\big)\Big|_{z=0} = \sum_{\mu=0}^{m} \binom{h}{j-\mu}(j-\mu)!F_{m-\mu,j-\mu}$$

where terms in the sum on the right-hand side with $\mu > j$ are put equal to $0$. All these expressions evidently are polynomials in $h$ with coefficients in $K$.

In particular, for $j = t$ and hence $k = h + m - t$, the polynomial

$$F^{(h,k)}\big((f)\big)\Big|_{z=0} = \binom{h}{t}t!F_{m,t} + \sum_{\mu=1}^{m} \binom{h}{t-\mu}(t-\mu)!F_{m-\mu,t-\mu}$$

is by (27) of the exact degree $t$ in $h$ and certainly does not vanish identically. Hence there exists a smallest integer $s$ satisfying

$$0 \leq s \leq t$$

such that also the polynomial in $h$,

(29): $$F^{(h,k)}\big((f)\big)\Big|_{z=0}, \text{ where } k = h + m - s,$$

is *not identically zero*, while for $j = 0, 1, \ldots, s-1$ the polynomials (28) vanish identically. With this choice of $s$, $k$ will from now on always be defined by

(30): $$k = h + m - s ;$$

hence $k$ differs from $h$ only by a constant integer which may be positive, negative, or zero.

It follows now, firstly, that the left-hand side of the equation (24) does not involve any of the coefficients

$$f_{k+1}, f_{k+2}, \ldots, f_{k+s},$$

but depends only on the coefficients

$$f_0, f_1, \ldots, f_k,$$

and that, secondly, this left-hand side is linear in $f_k$ with the factor

(31): $$k!F^{(h,k)}\big((f)\big)\Big|_{z=0} = k! \sum_{\mu=0}^{m} \binom{h}{s-\mu}(s-\mu)!F_{m-\mu,s-\mu}.$$

By what has already been said, this factor is not identically zero as function of $k$ and hence is distinct from zero as soon as $k$ is sufficiently large.

**127.** From the results just obtained we derive now a recursive formula for the coefficients $f_k$. For this purpose put, firstly,

(32): $$A(k) = \begin{cases} \frac{k!}{h!} \\ \\ 1 \end{cases} \quad \text{and} \quad B(k) = \begin{cases} 1 & \text{if } k \geq h , \\ \\ \frac{h!}{k!} & \text{if } k \leq h ; \end{cases}$$

secondly, put

(33): $$a(k) = A(k)F^{(h,k)}((f))\Big|_{s=0} = A(k) \sum_{\mu=0}^{m} \binom{h}{s-\mu}(s-\mu)!F_{m-\mu,s-\mu} ;$$

and, thirdly, denote by $\phi_k = \phi_k(f_0,f_1,\ldots,f_{k-1})$ the double sum

(34): $$\phi_k = -B(k) \sum_{(\kappa)} \sum_{[\lambda]}^{*} \frac{P_{(\kappa)}^{(\lambda_0)}(0)}{\lambda_0!} \frac{(\kappa_1+\lambda_1)!}{\lambda_1!} \cdots \frac{(\kappa_N+\lambda_N)!}{\lambda_N!} f_{\kappa_1+\lambda_1} \cdots f_{\kappa_N+\lambda_N} ,$$

where the asterisk on the double sum signifies that all terms having one of the factors

$$f_k, f_{k+1}, \ldots, f_{k+s}$$

are to be omitted.

From these definitions, $A(k) \not\equiv 0$ and $B(k) \not\equiv 0$ are polynomials in $k$ with rational integral coefficients; $a(k) \not\equiv 0$ is a polynomial in $k$ with coefficients in $K$ ; and $\phi_k$ is a polynomial in $f_0, f_1, \ldots, f_{k-1}$ with coefficients in $K$ .

The equation (24) leads immediately to the basic recursive formula

(35): $$a(k)f_k = \phi_k(f_0,f_1,\ldots,f_{k-1}) .$$

Here, since $a(k) \not\equiv 0$ , there exists a positive integer $k_0$ such that

(36): $$a(k) \neq 0 \quad \text{if} \quad k \geq k_0 .$$

Thus, as soon as $k \geq k_0$ , (35) expresses $f_k$ as a polynomial in the preceding coefficients $f_0, f_1, \ldots, f_{k-1}$ with coefficients in $K$ .

From this representation, we shall in the next sections deduce upper estimates for the absolute values of the coefficients $f_k$ ; and in the special case when these coefficients are algebraic numbers, we shall further establish a lower bound for their absolute values whenever they are distinct from zero.

**128.** For the moment, let $\kappa_0, \kappa_1, \ldots, \kappa_N$ be arbitrary non-negative integers, and let

$$w_\nu = \kappa_\nu!(1-z)^{-(\kappa_\nu+1)} \qquad (\nu = 0,1,\ldots,N) ,$$

so that

$$w_\nu^{(\lambda_\nu)} = (\kappa_\nu+\lambda_\nu)!(1-z)^{-(\kappa_\nu+\lambda_\nu+1)} \qquad (\nu = 0,1,\ldots,N) ,$$

and also

$$\frac{1}{h!}\left(\frac{d}{dz}\right)^h (w_0 w_1 \ldots w_N) = \kappa_0!\kappa_1! \ldots \kappa_N! \binom{\kappa_0+\kappa_1+\ldots+\kappa_N+h+N}{\kappa_0+\kappa_1+\ldots+\kappa_N+N} (1-z)^{-(\kappa_0+\kappa_1+\ldots+\kappa_N+h+N+1)} .$$

On the other hand, by (6),

$$\frac{1}{h!}\left(\frac{d}{dz}\right)^h (w_0 w_1 \ldots w_N) = \sum_{\lambda_0,\lambda_1,\ldots,\lambda_N} \frac{(\kappa_0+\lambda_0)!}{\lambda_0!} \ldots \frac{(\kappa_N+\lambda_N)!}{\lambda_N!} (1-z)^{-(\kappa_0+\kappa_1+\ldots+\kappa_N+h+N+1)} ,$$

where the summation extends again over all ordered systems $[\lambda] = [\lambda_0,\lambda_1,\ldots,\lambda_N]$ of integers satisfying

(7): $$\lambda_0 \geq 0,\ \lambda_1 \geq 0,\ \ldots,\ \lambda_N \geq 0;\ \lambda_0 + \lambda_1 + \ldots + \lambda_N = h .$$

On comparing these two formulae, it follows that

(37): $$\sum_{[\lambda]} \frac{(\kappa_0+\lambda_0)!}{\lambda_0!} \frac{(\kappa_1+\lambda_1)!}{\lambda_1!} \ldots \frac{(\kappa_N+\lambda_N)!}{\lambda_N!} = \binom{\kappa_0+\kappa_1+\ldots+\kappa_N+h+N}{\kappa_0+\kappa_1+\ldots+\kappa_N+N} \kappa_0!\kappa_1! \ldots \kappa_N! .$$

In particular, let

$$\kappa_0 = 0;\ 0 \leq \kappa_1 \leq m,\ \ldots,\ 0 \leq \kappa_N \leq m;\ 1 \leq N \leq n .$$

Then

$$0 \leq \kappa_0 + \kappa_1 + \ldots + \kappa_N + N \leq (m+1)n ,$$

so that trivially

$$\binom{\kappa_0+\kappa_1+\ldots+\kappa_N+h+N}{\kappa_0+\kappa_1+\ldots+\kappa_N+N} \leq \{h+(m+1)n\}^{(m+1)n} .$$

The equation (37) implies therefore in the present case that

(38): $$\sum_{[\lambda]} \frac{(\kappa_1+\lambda_1)!}{\lambda_1!} \ldots \frac{(\kappa_N+\lambda_N)!}{\lambda_N!} \leq m^{mn}\{h+(m+1)n\}^{(m+1)n} .$$

**129.** There are only finitely many polynomials $p_{(\kappa)}(z)$, and these have only finitely many non-zero coefficients

$$\frac{P_{(\kappa)}^{(\lambda_0)}(0)}{\lambda_0!}\ .$$

The maximum

$$c_0 = \max_{(\kappa),[\lambda]} \left|\frac{P_{(\kappa)}^{(\lambda_0)}(0)}{\lambda_0!}\right|$$

of the absolute values of all these coefficients is then a positive constant which is independent of $k$ .

In the formula (34), the double sum $\sum\limits_{(\kappa)}\sum\limits_{[\lambda]}^{*}$ is a subsum of the double sum $\sum\limits_{(\kappa)}\sum\limits_{[\lambda]}$ . Hence, by (30), (34), (35), and (38),

(39): $|a(k)||f_k| \le |B(k)|.c_0.m^{mn}\{k+(m+1)n-m+s\}^{(m+1)n} \ . \max^*_{(\kappa),[\lambda]} |f_{\kappa_1+\lambda_1}\cdots f_{\kappa_N+\lambda_N}|$ ,

where the maximum is extended over all pairs of systems $(\kappa)$, $[\lambda]$ for which

(40): $1 \le N \le n;\ 0 \le \kappa_1+\lambda_1 \le k-1,\ \ldots,\ 0 \le \kappa_N+\lambda_N \le k-1$ .

This estimate can be replaced by a slightly simpler one. Let $k_0$ be the same positive integer as in (36). There exist two further positive integers $c_1$ and $c_2$ , independent of $k$ , such that

$$\left|\frac{B(k)}{a(k)}\right| \le k^{c_1} \quad \text{for} \quad k \ge k_0\ ,$$

and therefore also

(41): $\left|\frac{B(k)}{a(k)}\right|.c_0.m^{mn}\{k+(m+1)n-m+s\}^{(m+1)n} \le k^{c_2}$ for $k \ge k_0$ .

Next, if $(\kappa)$ and $[\lambda]$ are any two systems as in (39) and (40), define a new ordered system $\{\nu\} = \{\nu_1,\ldots,\nu_N\}$ by

(42): $\nu_1 = \kappa_1+\lambda_1,\ \ldots,\ \nu_N = \kappa_N + \lambda_N$ .

Then, firstly, by (40),

(43): $1 \le N \le n;\ 0 \le \nu_1 \le k-1,\ \ldots,\ 0 \le \nu_N \le k-1$ .

Secondly, by the properties of $(\kappa)$ and $[\lambda]$ ,

$$\nu_1 + \ldots + \nu_N = (\kappa_1+\ldots+\kappa_N) + h = k + (\kappa_1+\ldots+\kappa_N-m+s)\ ,$$

and hence there exists a further positive integer $c_3$ independent of $k$ such that

(44): $$\nu_1 + \ldots + \nu_N \le k + c_3 .$$

We deduce finally from (39), (41), and the definition of $\{\nu\}$ that

(45): $$|f_k| \le k^{c_2} \cdot \max_{\{\nu\}} |f_{\nu_1} \ldots f_{\nu_N}| \quad \text{for} \quad k \ge k_0 ,$$

where the maximum is extended over all systems $\{\nu\}$ with the properties (43) and (44).

**130.** An upper bound for $|f_k|$ is now obtained in the following way from the recursive inequality (45).

Assume for simplicity that

(46): $$k_0 > c_3 + 1 .$$

Further choose $k_0$ numbers $u_0, u_1, \ldots, u_{k_0-1}$ such that

(47): $$|f_k| \le e^{u_k} \quad \text{for} \quad k = 0, 1, \ldots, k_0-1 ; \quad 0 < u_0 < u_1 < \ldots < u_{k_0-1} ,$$

and in terms of these define for each suffix $k \ge k_0$ a number $u_k$ by the equation

(48): $$u_k = c_2 \log k + \max_{\{\nu\} \in S_k} \left(u_{\nu_1} + \ldots + u_{\nu_N}\right) .$$

Here $S_k$ denotes the set of all ordered systems $\{\nu\} = \{\nu_1, \ldots, \nu_N\}$ of integers for which

(43-44): $1 \le N \le n$ ; $0 \le \nu_1 \le k-1, \ldots, 0 \le \nu_N \le k-1$ ; $\nu_1 + \ldots + \nu_N \le k + c_3$ ;

$S_k$ is thus the set of all systems $\{\nu\}$ over which the maximum in (45) is extended.

The definition of $u_k$ implies that

(49): $$|f_k| \le e^{u_k} \quad \text{for all suffixes} \quad k \ge 0 .$$

This is evident for $k \le k_0-1$, and it follows for $k \ge k_0$ by complete induction on $k$, since by (45) and (48),

$$|f_k| \le \exp\left(c_2 \log k + \max_{\{\nu\} \in S_k} \left(u_{\nu_1} + \ldots + u_{\nu_N}\right)\right) = e^{u_k} .$$

Again suppose $k \ge k_0$, so that by (46),

(50): $$k > c_3 + 1 .$$

By (48),

$$(51):\quad u_{k+1} - u_k = c_2 \log \frac{k+1}{k} + \max_{\{\nu'\}\in S_{k+1}} \left(u_{\nu'_1}+\ldots+u_{\nu'_{N'}}\right) - \max_{\{\nu\}\in S_k} \left(u_{\nu_1}+\ldots+u_{\nu_N}\right) .$$

Here, by (43-44), every system in $S_k$ belongs also to $S_{k+1}$ . The maximum over $S_{k+1}$ is therefore not less than that over $S_k$ , and so (51) implies that

$$(52):\qquad u_{k+1} - u_k \geq c_2 \log \frac{k+1}{k} > 0 \quad \text{for} \quad k \geq k_0 .$$

On combining this with (47), we see that the $u_k$ form a *strictly increasing sequence of positive numbers.*

131. Consider now any system $\{\pi\} = \{\pi_1, \ldots, \pi_{N*}\}$ in $S_{k+1}$ at which the maximum

$$(53):\qquad \max_{\{\nu'\}\in S_{k+1}} \left(u_{\nu'_1}+\ldots+u_{\nu'_{N'}}\right) = u_{\pi_1} + \ldots + u_{\pi_{N*}}$$

is attained. Since the numbers $u_k$ are positive and form an increasing sequence, the suffixes $\pi_1, \ldots, \pi_{N*}$ cannot all be zero; and since further

$$\pi_1 + \ldots + \pi_{N*} \leq k + c_3 + 1 \quad \text{and} \quad k > c_3 + 1 ,$$

at most one of these suffixes can be as large as $k$ . Let, say,

$$\pi_{N*} > 0$$

be the largest of the suffixes $\pi_1, \ldots, \pi_{N*}$ (or one of them if several have this same maximum value); then the other suffixes

$$\pi_1, \ldots, \pi_{N*-1}$$

are non-negative and less than $k$ . Hence the system $\{\nu^0\} = \{\nu_1^0, \ldots, \nu_{N^0}^0\}$ defined by

$$N^0 = N* \; ; \quad \nu_1^0 = \pi_1, \ldots, \nu_{N^0-1}^0 = \pi_{N*-1} , \quad \nu_{N^0}^0 = \pi_{N*} - 1 \geq 0$$

belongs to the set $S_k$ , and hence it follows that

$$\max_{\{\nu\}\in S_k} \left(u_{\nu_1}+\ldots+u_{\nu_N}\right) \geq u_{\pi_1} + \ldots + u_{\pi_{N*-1}} + u_{\pi_{N*}-1} =$$

$$= \max_{\{\nu'\}\in S_{k+1}} \left(u_{\nu'_1}+\ldots+u_{\nu'_{N'}}\right) - \left(u_{\pi_{N*}} - u_{\pi_{N*}-1}\right) .$$

The formula (51) leads therefore to the inequality

(54): $$u_{k+1} - u_k \leq c_2 \log \frac{k+1}{k} + \max_{l=0,1,\ldots,k-1} (u_{l+1}-u_l) \quad \text{for} \quad k \geq k_0 .$$

Put

$$v_l = u_{l+1} - u_l \quad \text{and} \quad c_4 = \max\left(v_0, v_1, \ldots, v_{k_0-1}\right) ,$$

so that $c_4$ is a positive constant. By (54),

$$v_k \leq c_2 \log \frac{k+1}{k} + \max_{l=0,1,\ldots,k-1} v_l \quad \text{for} \quad k \geq k_0 ,$$

that is,

$$v_k \leq c_2 \log \frac{k+1}{k} + \max\left(c_4, v_{k_0}, v_{k_0+1}, \ldots, v_{k-1}\right) \quad \text{for} \quad k \geq k_0 .$$

We assert that also

(55): $$v_k \leq c_2 \log \frac{k+1}{k_0} + c_4 \quad \text{for} \quad k \geq k_0 .$$

For this inequality certainly holds if $k = k_0$ ; assume then that $k > k_0$ , and that the inequality has already been proved for all suffixes up to and including $k - 1$ . Then

$$\max\left(c_4, v_{k_0}, v_{k_0+1}, \ldots, v_{k-1}\right) \leq c_2 \log \frac{k}{k_0} + c_4 ,$$

whence also

$$v_k \leq c_2 \log \frac{k+1}{k} + \left(c_2 \log \frac{k}{k_0} + c_4\right) = c_2 \log \frac{k+1}{k_0} + c_4 ,$$

showing that (55) holds also for the suffix $k$ and therefore is always true.

132. On putting

$$c_5 = c_4 - c_2 \log k_0 ,$$

(55) can be written in the form

$$u_{k+1} - u_k \leq c_2 \log(k+1) + c_5 \quad \text{for} \quad k \geq k_0 .$$

We apply this inequality successively for the suffixes $k_0$, $k_0+1$, ..., $k-1$ and add all the results, obtaining

$$u_k \leq u_{k_0} + c_2 \log(k!/k_0!) + c_5(k-k_0) \quad \text{for} \quad k \geq k_0 .$$

By (49), this finally implies that

(56): $$|f_k| \le e^{u_{k_0}+c_5(k-k_0)} (k!/k_0!)^{c_2} \quad \text{for} \quad k \ge k_0 .$$

In this formula, $k!$ increases more rapidly than any exponential function of $k$. On replacing $k$ again by $h$, we have then obtained the following estimate.

**THEOREM 16.** *Let*

$$f = \sum_{h=0}^{\infty} f_h z^h$$

*be a formal power series with real or complex coefficients which satisfies an algebraic differential equation. Then there exist two positive constants* $\gamma_1$ *and* $\gamma_2$ *such that*

$$|f_h| \le \gamma_1 (h!)^{\gamma_2} \qquad (h = 0,1,2,\ldots) .$$

In this proof I made use of an idea by a young Canberra mathematician, Dr A.N. Stokes, as contained in §130. See also Maillet (1903).

**133.** Of particular interest is one special case of Theorem 16, namely, when the coefficients $f_h$ of $f$ are (real or complex) *algebraic* numbers.

Let $K$ be the field obtained by adjoining all the coefficients $f_h$ to the rational field $Q$. By what was said in §120, we may then assume that also all the coefficients of the algebraic differential equation for $f$,

$$F((f)) = 0 ,$$

which are *finite* in number, lie in $K$. However, by the recursive formulae (35), the $f_h$ can be expressed rationally in terms of only *finitely many* of these numbers, and the coefficients of these relations involve only rational numbers and the coefficients of the differential polynomial $F((w))$. It follows therefore that $K$ can be obtained from $Q$ by adjoining only *finitely many algebraic numbers*, and hence $K$ is an algebraic number field of finite degree, $d$ day, over $Q$.

Denote by

$$K^{[0]} = K,\ K^{[1]},\ \ldots,\ K^{[d-1]}$$

the conjugate fields of $K$ over $Q$; if $\alpha$ is any element of $K$, let

$$\alpha^{[\delta]} \qquad (\delta = 0,1,\ldots,d-1)$$

be the conjugate of $\alpha$ in $K^{[\delta]}$; and as usual let

$$\overline{|\alpha|} = \max_{\delta=0,1,\ldots,d-1} |\alpha^{[\delta]}| .$$

We denote by

$$f^{[\delta]} = \sum_{h=0}^{\infty} f_h^{[\delta]} z^h \qquad (\delta = 0,1,\ldots,d-1)$$

the series in $K^{[\delta]}$ which is obtained from the series $f$ in $K$ when all its coefficients $f_h$ are replaced by their conjugates $f_h^{[\delta]}$ in $K^{[\delta]}$. Similarly, let

$$F^{[\delta]}\big((w)\big) = F^{[\delta]}(z;w,w',\ldots,w^{(m)})$$

be the differential polynomial which is obtained from $F\big((w)\big)$ when all its coefficients are replaced by their conjugates in $K^{[\delta]}$.

With this notation, it is evident that also

$$F^{[\delta]}\big((f^{[\delta]})\big) = 0 \qquad (\delta = 0,1,\ldots,d-1) .$$

Hence each of the series $f^{[\delta]}$ satisfies the hypothesis of Theorem 16, and hence there exist $2d$ positive constants

$$\gamma_1^{[\delta]} \text{ and } \gamma_2^{[\delta]} \qquad (\delta = 0,1,\ldots,d-1)$$

such that, for $\delta = 0, 1, \ldots, d-1$,

$$|f_h^{[\delta]}| \le \gamma_1^{[\delta]}(h!)^{\gamma_2^{[\delta]}} \qquad (h = 0,1,2,\ldots) .$$

On putting

$$\Gamma_1 = \max_{\delta=0,1,\ldots,d-1} \gamma_1^{[\delta]} \text{ and } \Gamma_2 = \max_{\delta=0,1,\ldots,d-1} \gamma_2^{[\delta]} ,$$

we have obtained the following result.

**THEOREM 17.** *Let*

$$f = \sum_{h=0}^{\infty} f_h z^h$$

*be a formal power series with (real or complex) algebraic coefficients which satisfies an algebraic differential equation. Then there exist two positive constants* $\Gamma_1$ *and* $\Gamma_2$ *such that*

$$\overline{|f_h|} \le \Gamma_1(h!)^{\Gamma_2} \qquad (h = 0,1,2,\ldots) .$$

**134.** We continue with the case when the coefficients $f_h$ of $f$ are algebraic numbers, hence when the results of the last section may be applied.

In terms of the earlier notation, it had been shown that

$$(35): \qquad a(k)f_k = \phi_k(f_0,f_1,\ldots,f_{k-1}) \quad \text{for} \quad k \geq k_0 ,$$

where

$$(34): \quad \phi_k = -B(k) \sum_{(\kappa)} \sum_{[\lambda]}{}^{*} \frac{P_{(\kappa)}^{(\lambda_0)}(0)}{\lambda_0!} \frac{(\kappa_1+\lambda_1)!}{\lambda_1!} \cdots \frac{(\kappa_N+\lambda_N)!}{\lambda_N!} f_{\kappa_1+\lambda_1} \cdots f_{\kappa_N+\lambda_N} .$$

Now, for each $k \geq 0$, denote by $d_k$ a suitable positive rational integer for which the product

$$(57): \qquad g_k = d_k f_k \qquad (k = 0,1,2,\ldots)$$

is an algebraic integer in $K$. Our next aim will be to find an upper estimate for how large such factors $d_k$ have to be.

The two sides of (35) may still be multiplied by an arbitrary polynomial in $k$ with coefficients in $K$. This factor can evidently be selected in such a way that the following three properties hold, where we have again written $a(k)$ and $B(k)$ for the new polynomial factors.

(i): The polynomial $a(k)$ has *rational integral* coefficients and does not vanish for $k \geq k_0$, where $k_0$ has the same meaning as before.

(ii): The polynomial $B(k)$ has *algebraic integral* coefficients in $K$.

(iii): All the values $\dfrac{P_{(\kappa)}^{(\lambda_0)}(0)}{\lambda_0!}$ are *algebraic integers* in $K$.

It follows then from (34) and (35) that

$$(58): \qquad a(k)f_k = \sum_{\{\nu\}\in S_k} P_{\{\nu\}}(k) f_{\nu_1} \cdots f_{\nu_N} \quad \text{for} \quad k \geq k_0 .$$

Here $\{\nu\}$ and $S_k$ are defined just as in §130, and the new coefficients $P_{\{\nu\}}(k)$ are integers in $K$ depending on both $\{\nu\}$ and $k$, but the exact values of which are for our purpose immaterial.

On account of (57), (58) is equivalent to

$$(59): \qquad g_k = \sum_{\{\nu\}\in S_k} P_{\{\nu\}}(k) \frac{d_k}{a(k) d_{\nu_1} \cdots d_{\nu_N}} g_{\nu_1} \cdots g_{\nu_N} \quad \text{for} \quad k \geq k_0 .$$

**135.** Choose now the first $k_0 + 1$ positive integers

$$d_0,\ d_1,\ \dots,\ d_{k_0}$$

fixed once for all such that

$$g_0 = d_0 f_0\ ,\quad g_1 = d_1 f_1,\ \dots,\ g_{k_0} = d_{k_0} f_{k_0}$$

are algebraic integers, and further select positive integers $d_k$ recursively such that

(60): $$a(k) d_{\nu_1} \dots d_{\nu_N} \mid d_k \quad \text{for all} \quad \{\nu\} \in S_k \quad \text{and all} \quad k \geq k_0+1\ .$$

It is then evident, by induction on $k$ and by (59), that also all the products

$$g_k = d_k f_k\ , \text{ where } \ k \geq k_0+1\ ,$$

are algebraic integers.

The relations (60) can be simplified a little by means of the following notation.

If $\{\nu\} = \{\nu_1,\dots,\nu_N\}$ is any system in $S_k$, distribute the suffixes $\nu_1,\ \dots,\ \nu_N$ into two subsets

$$\{\xi_1,\dots,\xi_X\} \quad \text{and} \quad \{\zeta_1,\dots,\zeta_Y\}$$

according as they are $\leq k_0$, or $\geq k_0+1$, respectively. Further, put

$$\eta_1 = \zeta_1 - k_0,\ \dots,\ \eta_Y = \zeta_Y - k_0$$

so that the $\eta$'s are positive integers. This subdivision is denoted symbolically by

$$\{\nu\} = \{\xi|\eta\} = \{\xi_1,\dots,\xi_X|\eta_1,\dots,\eta_Y\}\ .$$

It is obvious from its definition that

$$0 \leq X \leq n\ ,\quad 0 \leq Y \leq n\ , \text{ but } \ 1 \leq X+Y = N \leq n\ .$$

Next, if $l$ is any positive integer, put

(61): $$D_l = d_{l+k_0}\ ,$$

and write

$$\{\eta\} \in \Sigma_l \quad \text{if} \quad \{\xi|\eta\} \in S_{l+k_0}\ .$$

If $\{\xi|\eta\} \in S_{l+k_0}$, in each partial product

$$d_{\xi_1} \dots d_{\xi_X}$$

both the factors $d_{\xi_i}$ and their number $X$ are bounded. Therefore, for all $l$ and for all $\{\xi|\eta\}$, these partial products assume only finitely many distinct values, and hence a positive integer $\Delta$ exists such that always

$$d_{\xi_1} \dots d_{\xi_X} | \Delta .$$

Now put

(62): $$\alpha(l) = \Delta . a(l+k_0) .$$

Also $\alpha(l)$ is a polynomial with rational integral coefficients, and by (36),

(63): $$\alpha(l) \neq 0 \quad \text{for} \quad l = 1, 2, 3, \dots .$$

It follows then that the conditions (60) are equivalent to

$$a(l+k_0) d_{\xi_1} \dots d_{\xi_X} D_{\eta_1} \dots D_{\eta_Y} | D_l \quad \text{for all} \quad \{\xi|\eta\} \in S_{l+k_0} \qquad (l = 1,2,3,\dots)$$

and so are certainly satisfied if

(64): $$\alpha(l) D_{\eta_1} \dots D_{\eta_Y} | D_l \quad \text{for all} \quad \{\eta\} \in \Sigma_l \qquad (l = 1,2,3,\dots) .$$

Here we had

(43-44): $$0 \leq \nu_1 \leq k-1, \; \dots, \; 0 \leq \nu_N \leq k-1, \; \nu_1 + \dots + \nu_N \leq k+c_3 .$$

Therefore also

(65): $$1 \leq \eta_1 \leq l-1, \; \dots, \; 1 \leq \eta_Y \leq l-1$$

and

$$(\eta_1+k_0) + \dots + (\eta_Y+k_0) \leq (l+k_0) + c_3 ,$$

whence

$$\eta_1 + \dots + \eta_Y \leq l - k_0(Y-1) + c_3 .$$

Here, by the earlier assumption,

$$k_0 \geq c_3+1 \text{ , and therefore } -k_0(Y-1) + c_3 \leq -1 \quad \text{if} \quad Y \geq 2 .$$

Since the cases $Y = 0$ and $Y = 1$ are trivial, it follows that always

(66): $$\eta_1 + \dots + \eta_Y \leq l-1 .$$

**136.** It suffices then to choose the integers $D_l$ so as to satisfy the conditions (64) for all systems of integers $\{\eta\} = \{\eta_1,\dots,\eta_Y\}$ with the properties (65), (66), and

(67): $$1 \leq Y \leq n .$$

For this purpose, it suffices in fact to take

$$(68):\qquad D_l = \prod_{L=1}^{l} |\alpha(L)|^{\left[\frac{(n-1)l+1}{(n-1)L+1}\right]} \qquad (l = 1,2,3,\ldots)\ ,$$

where for the moment, $[x]$ has its usual meaning, the integral part of $x$. For this formula (68) is certainly satisfactory when $l = 1$; take $l \geq 2$, and assume that (68) has already been verified for all suffixes less than $l$. The exponent

$$\left[\frac{(n-1)l+1}{(n-1)L+1}\right]$$

is equal to 1 if $L = l$, has a positive integral value if $1 \leq L \leq l-1$, and vanishes if $L > l$. It suffices therefore to show that

$$(69):\qquad \left[\frac{(n-1)\eta_1+1}{(n-1)L+1}\right] + \ldots + \left[\frac{(n-1)\eta_Y+1}{(n-1)L+1}\right] \leq \left[\frac{(n-1)l+1}{(n-1)L+1}\right]$$

for all systems $\{\eta\}$ with the properties (65), (66), and (67), and for all integers $L$ in the interval $1 \leq L \leq l$. But under these hypotheses about $\{\eta\}$ and $L$,

$$\big((n-1)\eta_1+1\big) + \ldots + \big((n-1)\eta_Y+1\big) \leq (n-1)(l-1) + Y \leq (n-1)l + (Y-n) + 1 \leq (n-1)l + 1\ ,$$

which proves that (69) does hold.

In particular, the differential equation $F\big((f)\big) = 0$ for $f$ is linear if $n = 1$, and then (68) implies the simpler formula

$$(70):\qquad D_l = \prod_{L=1}^{l} |\alpha(L)| \qquad (l = 1,2,3,\ldots)\ .$$

An upper bound for $D_l$ can now be obtained as follows. Since $\alpha(l)$ is a polynomial, there exist two positive constants $c_6$ and $c_7$ such that

$$|\alpha(l)| \leq c_6 l^{c_7} \qquad (l = 1,2,3,\ldots)\ .$$

There is further a positive constant $c_8$ such that

$$\left[\frac{(n-1)l+1}{(n-1)L+1}\right] \leq c_8 \frac{l}{L} \quad \text{for all} \quad l \geq 1 \quad \text{and} \quad 1 \leq L \leq l\ ,$$

and there are also two positive constants $c_9$ and $c_{10}$ such that

$$\sum_{L=1}^{l} \frac{1}{L} \leq c_9 + \log l \quad \text{and} \quad \sum_{L=1}^{l} \frac{\log L}{L} \leq c_{10} + \frac{1}{2}(\log l)^2 \quad \text{for} \quad l \geq 1\ .$$

Hence it follows from (68) that

$$(71):\qquad 1 \leq D_l \leq \prod_{L=1}^{l} \left(c_6 L^{c_7}\right)^{c_8 \frac{l}{L}} \leq c_6^{\,c_8 l(c_9+\log l)} \cdot e^{c_7 c_8 l\{c_{10}+\frac{1}{2}(\log l)^2\}}\ .$$

In the linear case $n = 1$, we may instead use (70) and then arrive at the better estimate

$$(72): \qquad 1 \le D_l \le \prod_{L=1}^{l} \left(c_6 L^{c_7}\right) \le c_6^l (l!)^{c_7} .$$

For $0 \le k \le k_0$ , $d_k$ has the upper bound $d$ , while $d_k = D_{k-k_0}$ for larger suffixes. Hence, on changing again to the suffix $h$ , (71) and (72) lead immediately to the following result.

**THEOREM 18.** *Let*

$$f = \sum_{h=0}^{\infty} f_h z^h$$

*be a formal power series with (real or complex) algebraic coefficients, which satisfies an algebraic differential equation. Let* $d_h$ *be the smallest positive rational integer such that, for each* $h$ , $d_h f_h$ *is an algebraic integer.*

*Then there exists a positive constant* $\Gamma_3$ *such that*

$$(73): \qquad 1 \le d_h \le e^{\Gamma_3 h(\log h)^2} \quad \textit{for all sufficiently large } h .$$

*If, in fact, the differential equation for* $f$ *is linear, then there is a positive constant* $\Gamma_4$ *such that*

$$(74): \qquad 1 \le d_h \le e^{\Gamma_4 h \log h} \quad \textit{for all sufficiently large } h .$$

**137.** Let the hypothesis still remain as in the last two theorems, and let $h$ be any suffix such that $f_h \neq 0$ . Then $d_h f_h$ is an algebraic integer not zero, and hence its norm satisfies the inequality

$$\prod_{\delta=0}^{d-1} \left| \left(d_h f_h^{[\delta]}\right) \right| \ge 1 .$$

Here Theorem 17 gives upper estimates for the factors

$$f_h^{[\delta]} \qquad (\delta = 0,1,\ldots,d-1) ,$$

while Theorem 18 gives such an estimate for $d_h$ . We obtain then finally the following lower bound.

**THEOREM 19** (Theorem of J. Popken). *Let*

$$f = \sum_{h=0}^{\infty} f_h z^h$$

*be a formal power series with (real or complex) algebraic coefficients which satisfies an algebraic differential equation. There exists a positive constant* $\Gamma_5$ *such that, for all sufficiently large* $h$ *,*

$$\text{either } f_h = 0 \text{ , or } |f_h| \geq e^{-\Gamma_5 h(\log h)^2} .$$

*If, in fact, the differential equation for* $f$ *is linear, then there is a positive constant* $\Gamma_6$ *such that, for all sufficiently large* $h$ *,*

$$\text{either } f_h = 0 \text{ , or } |f_h| \geq e^{-\Gamma_6 h.\log h} .$$

**138.** In the next sections, Popken's theorem will be used to prove a result on transcendency. This result is due to Th. Schneider (1934 and 1957) who obtained it by an entirely different method.

Let, as usual,

$$\sigma(u) = \sigma(u|2\omega,2\omega') = \sigma(u;g_2,g_3)$$

be the Weierstrass $\sigma$-function, and let

$$p(u) = p(u|2\omega,2\omega') = p(u;g_2,g_3) = -\frac{d^2\log\sigma(u)}{du^2}$$

be Weierstrass's elliptic $p$-function, with the periods $2\omega$ and $2\omega'$ and the rational invariants $g_2$ and $g_3$. Here the notation is such that the imaginary part of $\omega'/\omega$ is positive, hence that

$$q = e^{\pi i\omega'/\omega}$$

satisfies the inequality

$$0 < |q| < 1 .$$

We further denote by $\eta$ the quantity depending on $\omega$ and $\omega'$ , or equivalently on $g_2$ and $g_3$ , for which

$$\sigma(u+2\omega) = -e^{2\eta(u+\omega)}\sigma(u) .$$

Since

$$p'(u)^2 = 4p(u)^3 - g_2 p(u) - g_3 ,$$

$\sigma(u)$ is an integral of the algebraic differential equation

$$(w^2w'''-3ww'w''+2w'^3)^2 - 4(w'^2-ww'')^3 + g_2w^4(w'^2-ww'') + g_3w^6 = 0$$

which is homogeneous of dimension 6 in the four functions $w$, $w'$, $w''$, and $w'''$.

We note that if $t \neq 0$ is a parameter,

$$\sigma(tu|2t\omega,2t\omega') = \sigma\left(tu;t^{-4}g_2,t^{-6}g_3\right) = t\sigma(u;g_2,g_3) .$$

The same change from $u$, $2\omega$, $2\omega'$ to $tu$, $2t\omega$, $2t\omega'$, respectively has the effect of transforming $\eta$, $g_2$, and $g_3$ into $t^{-1}\eta$, $t^{-4}g_2$, and $t^{-6}g_3$, respectively.

It is known that $\sigma(u)$ is an entire transcendental odd function of $u$ and therefore for all values of $u$ can be written as a convergent power series

$$\sigma(u) = \sum_{n=0}^{\infty} a_n u^{2n+1} .$$

Here the first coefficients are

$$a_0 = 1 , \quad a_1 = 0 , \quad a_2 = -\frac{g_2}{240} , \quad a_3 = -\frac{g_3}{840} , \quad a_4 = -\frac{g_2^2}{161280} .$$

Generally, $a_n = a_n(g_2,g_3)$ is a polynomial in $g_2$ and $g_3$ with rational coefficients with the homogeneity property

$$a_n\left(t^{-4}g_2,t^{-6}g_3\right) = t^{-2n}a_n(g_2,g_3) .$$

Naturally, $a_n$ may also be considered as a polynomial in $a_2$ and $a_3$ with rational coefficients and with the analogous homogeneity property.

**139.** There are advantages in considering instead of $\sigma(u)$ the new function

$$s(u) = s(u|2\omega,2\omega') = s(u;g_2,g_3) = e^{-\frac{\eta u^2}{2\omega}}\sigma(u)$$

which is likewise entire and transcendental. Since

$$\frac{\sigma'(u)}{\sigma(u)} = \frac{s'(u)}{s(u)} + \frac{\eta}{\omega}u ,$$

$s(u)$ also satisfies an algebraic differential equation of order 3 . The explicit form of this differential equation will not be needed.

Next, $s(u)$ has the homogeneity property

$$s(tu|2t\omega,2t\omega') = s\left(tu;t^{-4}g_2,t^{-6}g_3\right) = ts(u;g_2,g_3) .$$

As an entire odd function, $s(u)$ has an everywhere convergent power series expansion,

$$s(u) = \sum_{n=0}^{\infty} b_n u^{2n+1} .$$

Here, by the connection between $s(u)$ and $\sigma(u)$ ,

$$b_n = \sum_{k=0}^{n} \frac{1}{k!}\left(-\frac{\eta}{2\omega}\right)^k a_{n-k} .$$

Hence the first coefficients are

$$b_0 = 1 , \quad b_1 = -\frac{\eta}{2\omega} , \quad b_2 = a_2 - \frac{1}{2}\left(\frac{\eta}{2\omega}\right)^2 = -\frac{g_2}{240} - \frac{\eta^2}{8\omega^2} ,$$

$$b_3 = a_3 - \frac{\eta}{2\omega} a_2 - \frac{1}{6}\left(\frac{\eta}{2\omega}\right)^3 = -\frac{g_3}{840} + \frac{g_2\eta}{480\omega} - \frac{\eta^3}{48\omega^3} ,$$

so that, conversely,

$$\frac{\eta}{\omega} = -2b_1 , \quad g_2 = -120\left(2b_2+b_1^2\right) , \quad g_3 = -280\left(3b_3-3b_2b_1-2b_1^3\right) .$$

Generally, $b_n = b_n\left(\frac{\eta}{\omega}, g_2, g_3\right)$ is a polynomial in $\frac{\eta}{\omega}$, $g_2$ , and $g_3$ with rational coefficients which evidently has the homogeneity property

$$b_n\left(t^{-2}\frac{\eta}{\omega}, t^{-4}g_2, t^{-6}g_3\right) = t^{-2n} b_n\left(\frac{\eta}{\omega}, g_2, g_3\right) .$$

Again we may instead consider $b_n$ as a polynomial with rational coefficients in $b_1$, $b_2$ , and $b_3$ , with the analogous homogeneity property.

**140.** The function $s(u)$ is essentially a theta function and is known to have the product representation

$$s(u) = \frac{2\omega}{\pi}\sin\frac{\pi u}{2\omega}\prod_{n=1}^{\infty}\left\{1 + \frac{4q^{2n}\sin^2\frac{\pi u}{2\omega}}{(1-q^{2n})^2}\right\} .$$

On replacing $u$ by the new variable

$$z = 2i\sin\frac{\pi u}{2\omega} ,$$

$s(u)$ may instead be written as

$$s(u) = \frac{\omega}{\pi i} f(z) ,$$

where $f(z)$ is defined by

$$f(z) = z \prod_{n=1}^{\infty} \left(1 - \frac{q^{2n} z^2}{(1-q^{2n})^2}\right) .$$

Again $f(z)$ is an entire transcendental function of $z$ and has an expansion

$$f(z) = \sum_{h=0}^{\infty} c_h z^{2h+1} .$$

Here $c_0 = 1$ , while for $h \geq 1$ ,

$$c_h = (-1)^h \sum_{n_1} \sum_{n_2} \cdots \sum_{n_h} \frac{q^{2(n_1+n_2+\ldots+n_h)}}{\{(1-q^{2n_1})(1-q^{2n_2})\ldots(1-q^{2n_h})\}^2} .$$

Here the summation extends over all ordered systems of $h$ integers $n_1, n_2, \ldots, n_h$ which satisfy the inequalities

$$1 \leq n_1 < n_2 < \ldots < n_h .$$

Now, since $0 < |q| < 1$ ,

$$|1-q^{2n}| \geq 1 - |q|^2 \qquad (n = 1,2,3,\ldots) .$$

Hence

$$|c_h| \leq \sum_{n_1=1}^{\infty} \sum_{n_2=2}^{\infty} \cdots \sum_{n_h=h}^{\infty} \frac{|q|^{2(n_1+n_2+\ldots+n_h)}}{(1-|q|^2)^{2h}} =$$

$$= \frac{1}{(1-|q|^2)^{2h}} \sum_{n_1=1}^{\infty} |q|^{2n_1} \sum_{n_2=2}^{\infty} |q|^{2n_2} \cdots \sum_{n_h=h}^{\infty} |q|^{2n_h} ,$$

whence

$$|c_h| \leq \frac{|q|^{h(h+1)}}{(1-|q|^2)^{3h}} .$$

**141.** Evidently,

$$\frac{ds(u)}{du} = \frac{1}{2} f'(z)(z^2+4)^{1/2} ,$$

with similar, but more complicated, expressions for the higher derivatives. The differential equation for the function $s(u)$ implies therefore that also the function $f(z)$ and, more generally, every section

$$\sum_{h=m}^{\infty} c_h z^{2h+1} \qquad (m = 1,2,3,\ldots)$$

of its power series satisfies an algebraic differential equation of the third order.

Moreover, all these sections are entire transcendental functions.

Therefore, by Popken's theorem and by the upper estimate for $|c_h|$, *infinitely many of the coefficients*

$$c_1, c_2, c_3, \ldots$$

*are transcendental numbers.*

These coefficients can be expressed in terms of the coefficients $b_n$ in the series

$$s(u) = \sum_{n=0}^{\infty} b_n u^{2n+1} .$$

For

$$u = \frac{2\omega}{\pi} \sin^{-1} \frac{z}{2i} ,$$

so that, by the classical power series for $\sin^{-1} x$,

$$u^{2n+1} = \left(\frac{\omega}{\pi i}\right)^{2n+1} \sum_{k=0}^{\infty} d_{n,k} z^{2k+2n+1} ,$$

where the coefficients $d_{n,k}$ are certain rational numbers. For the lowest values of $n$,

$$u = \frac{\omega}{\pi i} \left(z - \frac{z^3}{24} + \frac{3z^5}{640} - \frac{5.z^7}{7168} +-\ldots\right) ,$$

$$u^3 = \left(\frac{\omega}{\pi i}\right)^3 \left(z^3 - \frac{z^5}{8} + \frac{37z^7}{1920} -+\ldots\right) ,$$

$$u^5 = \left(\frac{\omega}{\pi i}\right)^5 \left(z^5 - \frac{5z^7}{24} +-\ldots\right) ,$$

$$u^7 = \left(\frac{\omega}{\pi i}\right)^7 (z^7 -+\ldots) .$$

By these developments,

$$s(u) = \sum_{n=0}^{\infty} b_n u^{2n+1} = \sum_{n=0}^{\infty} b_n \left(\frac{\omega}{\pi i}\right)^{2n+1} \sum_{k=0}^{\infty} d_{n,k} z^{2k+2n+1} = \frac{\omega}{\pi i} f(z) = \frac{\omega}{\pi i} \sum_{h=0}^{\infty} c_h z^{2h+1} ,$$

whence, on comparing the coefficients of equal powers of $z$,

$$c_h = \sum_{n=0}^{h} b_n \left(\frac{\omega}{\pi i}\right)^{2n} d_{n,h-n} \qquad (h = 1,2,3,\ldots) .$$

For the lowest values of $n$,

$$c_1 = \left(\frac{\omega}{\pi i}\right)^2 b_1 - \frac{1}{24} ,$$

$$c_2 = \left(\frac{\omega}{\pi i}\right)^4 b_2 - \frac{1}{8}\left(\frac{\omega}{\pi i}\right)^2 b_1 + \frac{3}{640} ,$$

$$c_3 = \left(\frac{\omega}{\pi i}\right)^6 b_3 - \frac{5}{24}\left(\frac{\omega}{\pi i}\right)^4 b_2 + \frac{37}{1920}\left(\frac{\omega}{\pi i}\right)^2 b_1 - \frac{5}{7186} .$$

Here

$$b_n = b_n\left(\frac{\eta}{\omega}, g_2, g_3\right) ,$$

hence

$$\left(\frac{\omega}{\pi i}\right)^{2n} b_n = b_n(G_1, G_2, G_3) , \quad = B_n \quad \text{say,}$$

where we have put

$$G_1 = \left(\frac{\omega}{\pi i}\right)^2 \frac{\eta}{\omega} = -\frac{\omega\eta}{\pi^2} , \quad G_2 = \left(\frac{\omega}{\pi i}\right)^4 g_2 = \frac{\omega^4}{\pi^4} g_2 , \quad G_3 = \left(\frac{\omega}{\pi i}\right)^6 g_3 = -\frac{\omega^6}{\pi^6} g_3 .$$

With this notation,

$$c_h = \sum_{n=0}^{h} d_{n,h-n} B_n \qquad (h = 1,2,3,\ldots) .$$

Hence all the coefficients $c_h$ are polynomials in $G_1$, $G_2$, and $G_3$ with rational coefficients.

Since the $c_h$ are not all algebraic, we thus arrive at the result that always at least one of the three numbers

$$G_1, \; G_2 , \text{ and } G_3$$

is transcendental. In particular, if $g_2$ and $g_3$ are both algebraic, then at least one of the two quotients

$$\frac{\omega}{\pi} \text{ and } \frac{\eta}{\pi}$$

is transcendental. This is weaker than Schneider's result that both numbers are transcendental.

We conclude this chapter with the conjecture that the exponent $\Gamma_5 h(\log h)^2$ in Popken's theorem can probably be improved to something like $\Gamma h(\log h)(\log\log h)$ .

# APPENDIX

## CLASSICAL PROOFS OF THE TRANSCENDENCY OF $e$ AND $\pi$

**1.** While J. Liouville in 1844 gave the first examples of transcendental numbers (Chapter 1, §10), the modern proofs of transcendency started with Hermite's paper of 1873. Already L. Euler had suggested that both $e$ and $\pi$ are transcendental, and the irrationality of $e^r$ for rational $r \neq 0$ and of $\pi$ had in fact been proved by J.H. Lambert in 1770. However, the transcendency of $e$ and $\pi$ was not established until a century later, that of $e$ by Hermite in the paper quoted, and that of $\pi$ (and more) by F. Lindemann in 1882. Lindemann obtained his results by a generalisation of Hermite's method, which thus proved basic for both problems.

Hermite's method depends on a set of formulae which approximate the ratios of several exponential functions by the ratios of the same number of polynomials. In this appendix, proofs of the transcendency of $e$ and $\pi$ and of Lindemann's more general results will be called *classical* if they are based on these formulae by Hermite. We shall here collect a number of such proofs by different mathematicians and explain their relations. Since Siegel's fundamental paper of 1929 (Chapters 4-7), entirely new and very powerful methods have been introduced into the theory of transcendental numbers. This has the danger that the earlier and often very ingenious classical proofs may be forgotten. For this reason it seems appropriate to look once again at the classical proofs.

**2.** Little arithmetic is required in Hermite's proof, just the fact that *the absolute value of a rational integer not zero is at least* 1 . Rather more is needed if one wishes to show Lindemann's results. Here, properties of algebraic numbers play a role. Let $\alpha^{(1)}, \alpha^{(2)}, \ldots, \alpha^{(d)}$ be a complete set of algebraic conjugates of an algebraic number. Then, firstly, *all symmetric functions with rational coefficients in these numbers are rational,* and secondly, *if* $\alpha^{(1)}$ *is an algebraic integer, then* $|\alpha^{(1)}\alpha^{(2)}\ldots\alpha^{(d)}|$ *cannot be less than* 1 *unless all the conjugates are equal to zero.*

In his paper, Lindemann proved the following

GENERAL LINDEMANN THEOREM. *Let* $\alpha_0, \alpha_1, \ldots, \alpha_n$ *be finitely many distinct algebraic numbers, and let* $a_0, a_1, \ldots, a_n$ *be an equal number of algebraic numbers, not all zero. Then the sum*

(1): $$a = a_0 e^{\alpha_0} + a_1 e^{\alpha_1} + \dots + a_n e^{\alpha_n}$$

*does not vanish.*

This theorem implies the transcendency of $e$ and, on account of Euler's equation

$$e^{\pi i} + 1 = 0 ,$$

also that of $\pi$. It further implies the transcendency of the elementary functions

$$e^x, \sin x, \tan x, \sin^{-1} x, \tan^{-1} x$$

for algebraic $x \neq 0$ and that of

$$\log x$$

for algebraic $x \neq 0$, $\neq 1$. It is easily seen to be equivalent to what was called the Hermite-Lindemann Theorem in §101.

3. K. Weierstrass (1885) was the first to prove that the General Lindemann Theorem is a consequence of the following

**SPECIAL LINDEMANN THEOREM.** *Let* $B = \{\beta_0, \beta_1, \dots, \beta_p\}$ *be a finite set of distinct algebraic numbers, containing all the conjugates of each of its elements. Let* $b_0, b_1, \dots, b_p$ *be non-zero rational integers such that* $b_{h_1} = b_{h_2} = \dots = b_{h_s}$ *whenever* $\beta_{h_1}, \beta_{h_2}, \dots, \beta_{h_s}$ *are algebraically conjugate. Then the sum*

(2): $$b = b_0 e^{\beta_0} + b_1 e^{\beta_1} + \dots + b_p e^{\beta_p}$$

*does not vanish.*

It suffices to prove this Special Lindemann Theorem because it implies the General Lindemann Theorem. For consider a sum (1) and assume that

$$a = 0 ,$$

contrary to the General Lindemann Theorem. Form the product

$$b = \prod_h \prod_k \left( a_0^{(h)} e^{\alpha_0^{(k)}} + a_1^{(h)} e^{\alpha_1^{(k)}} + \dots + a_n^{(h)} e^{\alpha_n^{(k)}} \right) = 0$$

extended over all the systems of algebraic conjugates of both the coefficients $a_0, a_1, \dots, a_n$ and the exponents $\alpha_0, \alpha_1, \dots, \alpha_n$. One proves easily that the product $b$ can be put in the form (2) with at least two non-zero terms. Since $b = 0$, the Special Lindemann Theorem would then also be false.

In the following considerations a number of proofs of both the General and the

Special Theorem of Lindemann will be given, as well as proofs of Hermite's less general result on the transcendency of $e$ .

# I

## Hermite's Approximation Functions

**4.** Denote by $m$ a positive integer and by

$$\Omega = (\omega_0,\omega_1,\ldots,\omega_m) \quad \text{and} \quad P = (\rho_0,\rho_1,\ldots,\rho_m)$$

two $(m+1)$-vectors, where the components of $\Omega$ are arbitrary real or complex numbers while those of $P$ are positive integer. Further put

$$(3): \qquad F(y) = \prod_{k=0}^{m} (y-\omega_k)^{\rho_k} \quad \text{and} \quad F(x;y) = \sum_{\lambda=0}^{\infty} x^{-\lambda-1} \frac{d^\lambda F(y)}{dy^\lambda} .$$

Since

$$\frac{d}{dy} e^y = e^y ,$$

$F(y)$ and $F(x;y)$ are connected by the integral formula

$$(4): \qquad \int F(y)e^{-xy}dy = -F(x;y)e^{-xy} .$$

Also put

$$(5): \qquad A_k\left(x\left|\begin{matrix}\omega_0\omega_1\ldots\omega_m\\ \rho_0\rho_1\ldots\rho_m\end{matrix}\right.\right) = A_k\left(x\left|\begin{matrix}\Omega\\ P\end{matrix}\right.\right) = x^{\rho_0+\rho_1+\ldots+\rho_m+1} F(x;\omega_k)$$

and

$$(6): \quad R_{kl}\left(x\left|\begin{matrix}\omega_0\omega_1\ldots\omega_m\\ \rho_0\rho_1\ldots\rho_m\end{matrix}\right.\right) = R_{kl}\left(x\left|\begin{matrix}\Omega\\ P\end{matrix}\right.\right) = x^{\rho_0+\rho_1+\ldots+\rho_m+1} e^{(\omega_k+\omega_l)x} \int_{\omega_l}^{\omega_k} F(y)e^{-xy}dy .$$

It follows from (4) that these functions satisfy Hermite's basic identities

$$(7): \qquad A_k\left(x\left|\begin{matrix}\Omega\\ P\end{matrix}\right.\right)e^{\omega_l x} - A_l\left(x\left|\begin{matrix}\Omega\\ P\end{matrix}\right.\right)e^{\omega_k x} = R_{kl}\left(x\left|\begin{matrix}\Omega\\ P\end{matrix}\right.\right) \qquad (k,l = 0,1,\ldots,m) .$$

Both kinds of functions $A_k$ and $R_{kl}$ have a number of remarkable properties which will be applied in the proofs of the theorems of Hermite and Lindemann.

**5.** From the definition of $F(y)$ , it follows easily that the derivative

$$\left.\frac{d^\lambda F(y)}{dy^\lambda}\right|_{y=\omega_k}$$

vanishes if

$$0 \le \lambda \le \rho_k - 1 \quad \text{or} \quad \lambda > \rho_0 + \rho_1 + \dots + \rho_m ,$$

but has the value

$$\rho_k! \prod_{\substack{h=0 \\ h\ne k}}^{m} (\omega_k - \omega_h)^{\rho_h} \quad \text{for} \quad \lambda = \rho_k ,$$

and the value

$$\rho_k! \left. \frac{d^{\lambda-\rho_k}}{dy^{\lambda-\rho_k}} \prod_{\substack{h=0 \\ h\ne k}}^{m} (y-\omega_h)^{\rho_h} \right|_{y=\omega_k} \quad \text{for} \quad \rho_k + 1 \le \lambda \le \rho_0 + \rho_1 + \dots + \rho_m .$$

Hence the definition (5) implies the following results.

(8): *As a function of* $x$, $A_k$ *is a polynomial of the form*

$$A_k\left(x \middle| \begin{smallmatrix}\Omega \\ \mathrm{P}\end{smallmatrix}\right) = x^{\sum_{h=0}^{m} \rho_h - \rho_k} \rho_k! \prod_{\substack{h=0 \\ h\ne k}}^{m} (\omega_k - \omega_h)^{\rho_h} \quad \textit{plus terms in lower powers of } x .$$

*Hence if* $\omega_0, \omega_1, \dots, \omega_m$ *are all distinct,* $A_k$ *has the exact degree*

$$\sum_{h=0}^{m} \rho_h - \rho_k .$$

(9): $A_k$ *is a polynomial also in* $\omega_0, \omega_1, \dots, \omega_m$, *and its homogeneous part in these variables of highest dimension has dimension not exceeding*

$$\sum_{h=0}^{m} \rho_h - \rho_k .$$

(10): $A_k$ *is a polynomial in the* $m+1$ *variables* $x, \omega_0, \omega_1, \dots, \omega_m$ *with rational integral coefficients divisible by* $\rho_k!$, *and it can be written as*

$$A_k\left(x \middle| \begin{smallmatrix}\Omega \\ \mathrm{P}\end{smallmatrix}\right) = x^{\sum_{h=0}^{m} \rho_h - \rho_k} \rho_k! \prod_{\substack{h=0 \\ h\ne k}}^{m} (\omega_k - \omega_h)^{\rho_h} + A_k^*\left(x \middle| \begin{smallmatrix}\Omega \\ \mathrm{P}\end{smallmatrix}\right)(\rho_k+1)!$$

*where also* $A_k^*$ *is a polynomial in* $x, \omega_0, \omega_1, \dots, \omega_m$ *with rational integral coefficients.*

6. Next, by simple results from function theory, the integral representation

(6) leads to the following results for $R_{kl}$ .

(11): $R_{kl}$ *is an entire transcendental function of* $x$ *which vanishes at* $x = 0$ *at least to the order* $\sum_{h=0}^{m} \rho_h + 1$ .

(12): *There exists a positive number* $c$ *which depends only on* $x, \omega_0, \omega_1, \ldots, \omega_m$ *such that*

$$|R_{kl}(x|{}^{\Omega}_{\mathrm{P}})| \leq c^{\sum_{h=0}^{m} \rho_h + 1} ,$$

*for all choices of the positive integers* $\rho_0, \rho_1, \ldots, \rho_m$ .

The last property (12) is particularly important because it is applied in most of the classical proofs of the transcendency of $e$ and $\pi$ .

It is possible to avoid integrals in the definition of the functions $R_{kl}$ . For this purpose one first introduces the polynomials $A_k$ as we have done, and one then defines $R_{kl}$ by means of the formula (7). Both A. Hurwitz (1893) and P. Gordan (1893) have given proofs of the inequality (12) based on this definition.

Hurwitz's proof assumes that all the variables $x, \omega_0, \omega_1, \ldots, \omega_m$ are real. One writes $R_{kl}$ in the form

$$R_{kl}(x|{}^{\Omega}_{\mathrm{P}}) = x^{\sum_{h=0}^{m} \rho_h + 1} e^{(\omega_k+\omega_l)x}\left(F(x;\omega_l)e^{-\omega_l x} - F(x;\omega_k)e^{-\omega_k x}\right)$$

and applies to the right hand side the mean value theorem of differential calculus. Since

$$\frac{d}{dy}\left(F(x;y)e^{-xy}\right) = -F(y)e^{-xy} ,$$

it follows that

$$R_{kl}(x|{}^{\Omega}_{\mathrm{P}}) = x^{\sum_{h=0}^{m} \rho_h + 1} e^{(\omega_k+\omega_l)x}(\omega_k - \omega_l)F(\omega_{kl})e^{-\omega_{kl}x}$$

where $\omega_{kl}$ is a certain number in the interval between $\omega_k$ and $\omega_l$ . A trivial estimate leads from this equation to the inequality (12).

Gordan's proof holds for arbitrary real or complex values of $x, \omega_0, \omega_1, \ldots, \omega_m$ . It is based on the equations

$$R_{kl}\left(x\middle|\begin{smallmatrix}\Omega\\ \mathrm{P}\end{smallmatrix}\right) = x^{\sum\limits_{h=0}^{m}\rho_h+1}\left\{F(x;\omega_l)e^{\omega_k x}-F(x;\omega_k)e^{\omega_l x}\right\} \quad\text{and}\quad F(x;y) = \sum_{\lambda=0}^{\infty} x^{-\lambda-1}\frac{d^{\lambda}F(y)}{dy^{\lambda}} .$$

Here the polynomial $F(y)$ can be written explicitly as

$$F(y) = \prod_{k=0}^{m}(y-\omega_k)^{\rho_k} = \sum_{r=0}^{\rho_0+\rho_1+\ldots+\rho_m} a_r y^r$$

where the $a_r$ are certain real or complex coefficients. A simple calculation leads to the development

$$R_{kl}\left(x\middle|\begin{smallmatrix}\Omega\\ \mathrm{P}\end{smallmatrix}\right) = x^{\sum\limits_{h=0}^{m}\rho_h+1}\sum_{r=0}^{\rho_0+\rho_1+\ldots+\rho_m} a_r x^{-r-1}\left\{e^{\omega_l x}\sum_{\lambda=r+1}^{\infty}\frac{r!}{\lambda!}(\omega_k x)^{\lambda}-e^{\omega_k x}\sum_{\lambda=r+1}^{\infty}\frac{r!}{\lambda!}(\omega_l x)^{\lambda}\right\} .$$

Here, by a trivial estimate,

$$\left|\sum_{\lambda=r+1}^{\infty}\frac{r!}{\lambda!}(xy)^{\lambda}\right| = \left|\frac{(xy)^{r+1}}{r+1}+\frac{(xy)^{r+2}}{(r+1)(r+2)}+\ldots\right| \le |xy|^{r+1}e^{|xy|} .$$

Thus if we put

$$F^*(y) = \sum_{r=0}^{\rho_0+\rho_1+\ldots+\rho_m}|a_r|y^r ,$$

it follows that

$$\left|R_{kl}\left(x\middle|\begin{smallmatrix}\Omega\\ \mathrm{P}\end{smallmatrix}\right)\right| \le |x|^{\sum\limits_{h=0}^{m}\rho_h+1}\left(|\omega_k|F^*(|\omega_k|)+|\omega_l|F^*(|\omega_l|)\right)e^{(|\omega_k|+|\omega_l|)|x|} ,$$

and from this formula an estimate (12) can be deduced without difficulty.

## II

## Application of the last Formulae to Proofs of Transcendency

7. In the application of the preceding formulae by Hermite to proofs of the transcendency of $e$ and $\pi$ it is usual to put $x = 1$. We then simply omit this argument and write

$$F[y],\ A_k\left(\begin{smallmatrix}\Omega\\ \mathrm{P}\end{smallmatrix}\right),\ R_{kl}\left(\begin{smallmatrix}\Omega\\ \mathrm{P}\end{smallmatrix}\right) \quad\text{for}\quad F(1;y),\ A_k\left(1\middle|\begin{smallmatrix}\Omega\\ \mathrm{P}\end{smallmatrix}\right),\ R_{kl}\left(1\middle|\begin{smallmatrix}\Omega\\ \mathrm{P}\end{smallmatrix}\right) ,$$

respectively. The second formula (3) and the formulae (5), (6), and (7) imply in this new notation that

(3'): $$F[y] = \sum_{\lambda=0}^{\infty} \frac{d^{\lambda}F(y)}{dy^{\lambda}} ,$$

(5'): $$A_k\binom{\Omega}{P} = F[\omega_k] ,$$

(6'): $$R_{kl}\binom{\Omega}{P} = e^{\omega_k+\omega_l} \int_{\omega_l}^{\omega_k} F[y]e^{-y}dy ,$$

(7'): $$A_l\binom{\Omega}{P}e^{\omega_k} - A_k\binom{\Omega}{P}e^{\omega_l} = R_{kl}\binom{\Omega}{P} \qquad (k,l = 0,1,\ldots,m) .$$

Let now $p$ be a positive integer not greater than $m$ , and let $o_0, o_1, \ldots, o_p$ be any $p + 1$ numbers. From (7') follows then the basic equation

(13): $$e^{-\omega_0}A_0\binom{\Omega}{P} \sum_{k=0}^{p} o_k e^{\omega_k} = \sum_{k=0}^{p} A_k\binom{\Omega}{P}o_k + e^{-\omega_0} \sum_{k=0}^{p} R_{k0}\binom{\Omega}{P}o_k .$$

In the applications of this formula, the parameters $\omega_0, \omega_1, \ldots, \omega_m$ and $o_0, o_1, \ldots, o_p$ will be fixed, while the positive integers $\rho_0, \rho_1, \ldots, \rho_m$ are variable. For the terms on the right-hand side (13) we shall write

(14): $$S = S(P) = S(\rho_0,\rho_1,\ldots,\rho_m) = \sum_{k=0}^{p} A_k\binom{\Omega}{P}o_k ,$$
$$T = T(P) = T(\rho_0,\rho_1,\ldots,\rho_m) = e^{-\omega_0} \sum_{k=0}^{p} R_{k0}\binom{\Omega}{P}o_k .$$

We further use the abbreviation

(15): $$\rho_{\min} = \min(\rho_0,\rho_1,\ldots,\rho_m) ,$$

(16): $$\sigma = \rho_0 + \rho_1 + \ldots + \rho_m .$$

From (9), (10), and (12) the following two properties follow at once.

(17): *$S$ is a polynomial in $\omega_0, \omega_1, \ldots, \omega_m$ and $o_0, o_1, \ldots, o_p$ with rational integral coefficients divisible by* $\rho_{\min}!$ .

(18): *There exists a positive constant $C$ which depends only on $\omega_0, \omega_1, \ldots, \omega_m$ and $o_0, o_1, \ldots, o_p$ such that*

$$|T| \leq C^{\sigma+1} .$$

8. Three different kinds of restrictions on the parameters $\omega_k$, $o_k$ , and $\rho_k$ will now be considered.

The simplest one is as follows.

HYPOTHESIS 1

(a): *The numbers* $\omega_0, \omega_1, \ldots, \omega_m$ *are rational integers, and in particular* $\omega_0, \omega_1, \ldots, \omega_p$ *are all distinct.*

(b): *The numbers* $c_0, c_1, \ldots, c_p$ *are rational integers distinct from zero.*

(c): *The numbers* $\rho_0, \rho_1, \ldots, \rho_m$ *are positive integers such that*

$$\rho_{\min}! > C^{\sigma+1} ,$$

*where* $C$ *is the constant occurring in* (18).

By (17), this hypothesis implies firstly that $S$ is a rational integer divisible by $\rho_{\min}!$ . Therefore

(19): $$S = 0 \quad \text{or} \quad |S| \geq \rho_{\min}! \; .$$

Secondly, $T$ satisfies the inequality (18).

Suppose now that $e$ were an algebraic number. Then $e$ satisfies an algebraic equation

$$c_0 e^{\omega_0} + c_1 e^{\omega_1} + \ldots + c_p e^{\omega_p} = 0$$

where $p \geq 1$ ; $c_0, c_1, \ldots, c_p$ are rational integers not zero; and $\omega_0, \omega_1, \ldots, \omega_p$ are distinct rational integers. Choose $m \geq p$ , and assume that $\omega_{p+1}, \omega_{p+2}, \ldots, \omega_m$ are arbitrary rational integers, while $\rho_0, \rho_1, \ldots, \rho_m$ are positive integers satisfying $\rho_{\min}! > C^{\rho+1}$ where $C > 0$ is the constant which by (18) belongs to $\omega_0, \omega_1, \ldots, \omega_m$ and $c_0, c_1, \ldots, c_p$ . Then, by the identity (13),

$$S + T = 0 \text{ , and hence also } |S| = |T| \; .$$

However, by the inequalities (18) and (19) and by the choice of the $\rho_k$ , this equation cannot hold unless simultaneously

$$S = 0 \quad \text{and} \quad T = 0 \; .$$

Hence, the following result holds.

THEOREM 1. *Assume that for all integers* $\omega_0, \omega_1, \ldots, \omega_m$ *and* $c_0, c_1, \ldots, c_p$ *with the properties* (a) *and* (b) *of Hypothesis* 1 *there exist integers* $\rho_0, \rho_1, \ldots, \rho_m$ *with the property* (c) *of Hypothesis* 1 *for which at least one of the two sums* $S(P)$ *and* $T(P)$ *does not vanish. Then* $e$ *is transcendental.*

9. We next establish two theorems both of which will enable us to prove the transcendency of $\pi$ and even the two theorems of Lindemann.

Denote by the letter $\Sigma$ (with suitable suffixes) sets which contain exactly *one algebraic number and all its algebraic conjugates.* Any two such sets $\Sigma_i$ and $\Sigma_j$ naturally either are identical, or are disjoint.

A second assumption on the parameters $\omega_k$, $o_k$, and $\rho_k$ can now be stated as follows.

HYPOTHESIS 2

(a): *The set* $M = \{\omega_0,\omega_1,\dots,\omega_m\}$ *of algebraic numbers contains a subset* $M^* = \{\omega_0,\omega_1,\dots,\omega_p\}$ *all elements of which are distinct; here* $1 \le p \le m$. *Further* $M$ *and* $M^*$ *can be split into the unions*

$$M = \bigcup_{j=1}^{u} \Sigma_j \quad \text{and} \quad M^* = \bigcup_{j=1}^{v} \Sigma_j$$

*where* $1 \le v \le u$, *and where the sets* $\Sigma_1, \Sigma_2, \dots, \Sigma_v$ *(but not necessarily also* $\Sigma_1, \Sigma_2, \dots, \Sigma_u$*) are all distinct and hence disjoint. Denote by* $N$ *a positive integer such that all the products* $N\omega_1, N\omega_2, \dots, N\omega_m$ *are algebraic integers.*

(b): *The numbers* $o_1, o_2, \dots, o_p$ *are rational integers not zero. Whenever* $1 \le k \le p$ *and* $1 \le l \le p$ *and* $\omega_k$ *and* $\omega_l$ *belong to the same set* $\Sigma_j$, *then* $o_k = o_l$.

(c): *The numbers* $\rho_0, \rho_1, \dots, \rho_m$ *are positive integers such that*

$$\rho_{\min}! > (CN)^{\sigma+1},$$

*where* $C$ *is the constant occurring in* (18). *Whenever* $\omega_k$ *and* $\omega_l$ *belong to the same set* $\Sigma_j$ *where* $1 \le j \le u$, *then* $\rho_k = \rho_l$.

These assumptions have a number of immediate consequences. Firstly, by (a),

(20): *the polynomial* $\prod_{k=0}^{m} (y-\omega_k)$ *has rational coefficients.*

Secondly, by (a) and (b),

(21): *If* $f(y)$ *is any polynomial with rational coefficients, then the sum*

$$\sum_{k=0}^{p} f(\omega_k)o_k$$

*has a rational value.*

Thirdly, by (a) and (c),

(22): *Both polynomials*

$$F(y) = \sum_{k=0}^{m} (y-\omega_k)^{\rho_k} \quad \text{and} \quad F[y] = \sum_{\lambda=0}^{\infty} \frac{d^\lambda F(y)}{dy^\lambda}$$

*have rational coefficients.*

It follows therefore from (21) that also the sum

$$S = \sum_{k=0}^{p} F[\omega_k]o_k$$

has a rational value. On the other hand, by (17), the quotient

$$\frac{S}{\rho_{\min}!}$$

is a polynomial in $\rho_0, \rho_1, \ldots, \rho_m$ with rational integral coefficients in which, by (9), only terms of dimension not exceeding

$$\max_{k=0,1,\ldots,m} \left(\sum_{h=0}^{m} \rho_h - \rho_k\right) < \sigma + 1$$

occur. The number

$$\frac{N^{\sigma+1}S}{\rho_{\min}!}$$

is therefore an algebraic integer, and being rational, is a *rational integer*. This implies that

$$S = 0 \quad \text{or} \quad |S| \geq \frac{\rho_{\min}!}{N^{\sigma+1}} .$$

On the other hand, by (18) and by the assumption (c),

$$|T| \leq C^{\sigma+1} < \frac{\rho_{\min}!}{N^{\sigma+1}} .$$

Assume now either that $\pi$ is algebraic, or more generally, that the Special Theorem and hence also the General Theorem of Lindemann are false. Then an equation

$$c_0 e^{\omega_0} + c_1 e^{\omega_1} + \ldots + c_p e^{\omega_p} = 0$$

holds where $p \geq 1$, and $c_0, c_1, \ldots, c_p$ and $\omega_0, \omega_1, \ldots, \omega_p$ are as in Hypothesis 2. Let $m \geq p$, and let $\omega_{p+1}, \omega_{p+2}, \ldots, \omega_m$ be $m - p$ further algebraic numbers, and let $\rho_0, \rho_1, \ldots, \rho_m$ be positive integers such that $\omega_0, \omega_1, \ldots, \omega_m$, $c_0, c_1, \ldots, c_p$, and $\rho_0, \rho_1, \ldots, \rho_m$ satisfy all three conditions (a), (b), and (c) of Hypothesis 2. Then again

$$S = T ,$$

and it follows just as in the case of the Hypothesis 1 that simultaneously

$$S = 0 \text{ and } T = 0 .$$

Hence the following result holds.

**THEOREM 2.** *Assume that for any algebraic numbers $\omega_0, \omega_1, \ldots, \omega_m$ and for any rational integers $c_0, c_1, \ldots, c_p$ with the properties (a) and (b) of Hypothesis 2 there exist positive integers $\rho_0, \rho_1, \ldots, \rho_m$ with the property (c) of Hypothesis 2 for which at least one of the two sums $S(P)$ and $T(P)$ does not vanish. Then both the Special Theorem and the General Theorem of Lindemann hold, and so in particular $\pi$ is transcendental.*

**10.** In order to formulate simply a further criterion for the transcendency of $\pi$, let $\delta_{hk}$ denote the Kronecker symbol which is defined by

$$\delta_{hh} = 1 , \quad \delta_{hk} = 0 \quad \text{if} \quad h \neq k .$$

The third assumption is then as follows.

HYPOTHESIS 3

(a): *The set $M = \{\omega_0, \omega_1, \ldots, \omega_m\}$ consists of different algebraic numbers and can be split into a union*

$$M = \bigcup_{j=1}^{u} \Sigma_j$$

*of the distinct and hence disjoint sets $\Sigma_1, \Sigma_2, \ldots, \Sigma_u$ defined as in §9. Denote by $N$ a positive integer such that all the products $N\omega_0, N\omega_1, \ldots, N\omega_m$ are algebraic integers.*

(b): *The numbers $c_0, c_1, \ldots, c_m$ are rational integers not zero. Whenever $\omega_k$ and $\omega_l$ belong to the same set $\Sigma_j$, then $c_k = c_l$.*

(c): *Let* $h$ *denote any one of the integers* $0, 1, \ldots, m$ *, and let* $\rho$ *be any positive integer satisfying*

$$(\rho-1)! > (CN)^{(m+1)\rho} ,$$

*where* $C$ *is the constant occurring in* (18). *Then define* $\rho_0, \rho_1, \ldots, \rho_m$ *by*

$$\rho_0 = \rho-\delta_{h0},\ \rho_1 = \rho-\delta_{h1},\ \ldots,\ \rho_m = \rho-\delta_{hm} .$$

Evidently, except for the specialisation $p = m$ , the assumptions (a) and (b) are the same as in Hypothesis 2. Therefore, by (20), the polynomial

$$\prod_{k=0}^{m} (y-\omega_k)^{\rho}$$

has rational coefficients, and so the quotients

$$F_h(y) = \prod_{k=0}^{m} (y-\omega_k)^{\rho-\delta_{hk}} = \frac{1}{y-\omega_h} \prod_{k=0}^{m} (y-\omega_k)^{\rho}$$

can be written in the form

$$F_h(y) = G(y,\omega_h)$$

where $G(y,z)$ is a certain polynomial in $y$ and $z$ with rational coefficients. Hence, on putting

$$F_h[y] = \sum_{\lambda=0}^{\infty} \frac{d^{\lambda}F_h(y)}{dy^{\lambda}} \quad \text{and} \quad G[y,z] = \sum_{\lambda=0}^{\infty} \frac{d^{\lambda}G(y,z)}{dy^{\lambda}} ,$$

it follows that

$$F_h[y] = G[y,\omega_h] .$$

Evidently also $G[y,z]$ is a polynomial in $y$ and $z$ with rational coefficients. Therefore, just as in (21), the sum

$$H(z) = \sum_{k=0}^{m} G[\omega_k,z]o_k$$

is a polynomial in $z$ with rational coefficients.

In terms of $F_h[y]$ and $G[y,z]$ ,

$$A_k\left(1\left|\begin{matrix}\omega_0 & \omega_1 & \cdots & \omega_m\\ \rho-\delta_{h0} & \rho-\delta_{h1} & \cdots & \rho-\delta_{hm}\end{matrix}\right.\right) = F_h[\omega_k] = G[\omega_k,\omega_h] .$$

The sum $S$ now depends on the choice of $h$ , and we write

$$S_h = S_h(\rho) = S(\rho-\delta_{h0},\rho-\delta_{h1},\ldots,\rho-\delta_{hm}) = \sum_{k=0}^{m} F_h[\omega_k]o_k = H(\omega_h) .$$

It follows therefore from (a) that all elements of the set

$$\{S_0,S_1,\ldots,S_m\}$$

are algebraic numbers, and with each of these numbers also all its algebraic conjugates belong to the set.

Next, by (17) and by the assumption (a), each of the quotients

$$\frac{S_h}{(\rho-1)!} \qquad (h = 0,1,\ldots,m)$$

is a polynomial with rational integral coefficients in $\omega_0, \omega_1, \ldots, \omega_m$ in which, by (9), only terms of dimension

$$\sum_{k=0}^{m} (\rho-\delta_{hk}) - (\rho-1) < (m+1)\rho$$

are present. Hence the set of numbers

$$\{s_0,s_1,\ldots,s_m\} \text{ , where } s_h = \frac{N^{(m+1)\rho}S_h}{(\rho-1)!} ,$$

consists of algebraic integers, and with each of its elements also all its algebraic conjugates belong to the set. The absolute value of the norm of each of these elements is then either 0 or not less than 1 , and hence the set of inequalities

$$|s_0| < 1,\ |s_1| < 1,\ \ldots,\ |s_m| < 1$$

can hold only if simultaneously

$$s_0 = s_1 = \ldots = s_m = 0 .$$

Therefore either

$$S_0 = S_1 = \ldots = S_m = 0 ,$$

or at least one of the inequalities

$$|S_h| \geq \frac{(\rho-1)!}{N^{(m+1)\rho}} \qquad (h = 0,1,\ldots,m)$$

holds.

Next, depending on the choice of $h$ , the sum $T$ becomes in this case

$$T_h = T_h(\rho) = T(\rho-\delta_{h0},\rho-\delta_{h1},\ldots,\rho-\delta_{hm}) = e^{-\omega_0} \sum_{k=0}^{m} R_{k0}\left(1\middle|\begin{matrix}\omega_0 & \omega_1 & \ldots & \omega_m\\ \rho-\delta_{h0} & \rho-\delta_{h1} & \ldots & \rho-\delta_{hm}\end{matrix}\right) .$$

By (18) and by the assumption (c) this expression satisfies the inequality

$$|T_h| \le C^{(m+1)\rho} < \frac{(\rho-1)!}{N^{(m+1)\rho}}$$

where we have used that in the present case

$$\sigma = (m+1)\rho - 1 .$$

Now assume again either that $\pi$ is algebraic, or more generally, that the Special Theorem and hence also the General Theorem of Lindemann are false. Then an equation

$$c_0 e^{\omega_0} + c_1 e^{\omega_1} + \dots + c_m e^{\omega_m} = 0$$

holds, in which the $\omega_k$ and $c_k$ satisfy the assumptions (a) and (b) of Hypothesis 3. Take for $\rho$ any positive integer for which also the assumption (c) of Hypothesis 3 is satisfied. By the same argument as before we now obtain a system of $m + 1$ equations

$$S_h = T_h \qquad (h = 0,1,\dots,m)$$

and, just as in the previous two cases, deduce from them that simultaneously

$$S_h = 0 \text{ and } T_h = 0 \qquad (h = 0,1,\dots,m) .$$

Hence we arrive at the following result.

**THEOREM 3.** *Assume that for any algebraic numbers* $\omega_0, \omega_1, \dots, \omega_m$ *and for any rational integers* $c_0, c_1, \dots, c_m$ *with the properties* (a) *and* (b) *of Hypothesis* 3 *there exist,*

*(i)* *a positive integer* $\rho$ *with the property* (c) *of Hypothesis* 3 *, and*

*(ii)* *a suffix* $h$ *in the interval* $0 \le h \le m$ *, such that at least one of the two sums* $S_h(\rho)$ *and* $T_h(\rho)$ *does not vanish.*

*Then both the Special Theorem and the General Theorem of Lindemann are true, and so in particular* $\pi$ *is transcendental.*

**11.** *The three theorems so established reduce the problems of the transcendency of* $e$ *and* $\pi$ *to that of deciding whether the sums* $S$ *or* $T$ *(or* $S_h$ *or* $T_h$ *) which belong to admissible sets of parameters* $\omega_k$, $c_k$ *, and* $\rho_k$ *(or* $\rho$ *) are equal to zero or not.*

As we shall see, we may distinguish between three classes of methods for making this decision, as follows.

(1): *Proofs in which it is shown that a certain determinant does not vanish. This method was first used by Hermite.*

(2): *Proofs which depend on divisibility properties. This method was introduced by Hilbert.*

(3): *Proofs based on analytic estimates. It was Stieltjes who first used this method.*

In the first two kinds of proof one establishes that sums $S$ or $S_h$ do not vanish, but in the third kind one proves instead that a certain sum $T$ is distinct from $0$.

## III

## Proofs depending on the Non-Vanishing of a Determinant

12. Let $1 \le p \le m$. For the moment $\omega_0, \omega_1, \ldots, \omega_m$ may be arbitrary, while $o_0, o_1, \ldots, o_p$ may be arbitrary numbers distinct from zero. Further, let

$$\rho_{h0}, \rho_{h1}, \ldots, \rho_{hm} \qquad (h = 0,1,\ldots,p)$$

be any $p+1$ distinct systems each consisting of $m+1$ positive integers.

*If the determinant of order* $p+1$,

$$D = \left| A_k \begin{pmatrix} \omega_0 & \omega_1 & \ldots & \omega_m \\ \rho_{h0} & \rho_{h1} & \ldots & \rho_{hk} \end{pmatrix} \right|_{h,k=0,1,\ldots,p}$$

*does not vanish, at least one of the* $p+1$ *sums*

$$S(\rho_{h0},\rho_{h1},\ldots,\rho_{hm}) = \sum_{k=0}^{p} A_k \begin{pmatrix} \omega_0 & \omega_1 & \ldots & \omega_m \\ \rho_{h0} & \rho_{h1} & \ldots & \rho_{hm} \end{pmatrix} o_k \qquad (h = 0,1,\ldots,p)$$

*is distinct from zero.*

This basic fact from linear algebra was applied by Hermite in two different ways to construct a determinant $D \neq 0$ and hence a sum $S \neq 0$; the transcendency of $e$ may then be derived from Theorem 1.

13. In his first proof, Hermite assumed that $\omega_0, \omega_1, \ldots, \omega_m$ are real and distinct, and without loss of generality arranged such that

$$\omega_0 < \omega_1 < \ldots < \omega_m .$$

Let $p = m$, and let the integers $\rho_{hk}$ have the form

$$\rho_{hk} = \rho + \theta_{hk} \qquad (h,\ k = 0,1,\ldots,m)$$

where $\rho$ denotes a positive integer tending to infinity, while the terms $\theta_{hk}$ are bounded rational integers which do not depend on $\rho$. Denote by $P_h$ the vectors

$$\mathrm{P}_h = (\rho_{h0}, \rho_{h1}, \ldots, \rho_{hm}) \qquad (h = 0,1,\ldots,m) .$$

Hermite uses Laplace's method to prove that asymptotically

$$A_m\binom{\Omega}{\mathrm{P}_h} \sim \left(\sum_{k=0}^{m} \rho_{hk}\right)! \exp\left(\frac{1}{m+1} \sum_{j=0}^{m} (\omega_k - \omega_j)\right) \qquad (h = 0,1,\ldots,m)$$

and

$$R_{k,k-1}\binom{\Omega}{\mathrm{P}_h} \sim e^{\omega_k - \omega_{k-1} - \zeta_k} \left(\frac{2\pi}{\rho\upsilon_k}\right)^{1/2} \tau_k^{\rho} \prod_{j=0}^{m} (\zeta_k - \omega_j)^{\theta_{hj}} \qquad \binom{h = 0,1,\ldots,m}{k = 1,2,\ldots,m} .$$

Here, in the second formula, $\zeta_k$ denotes the only root of the equation

$$\sum_{j=0}^{m} \frac{1}{\zeta - \omega_j} = 0$$

in the interval $\omega_{k-1} < \zeta < \omega_k$ , and $\tau_k$ and $\upsilon_k$ are defined by

$$\tau_k = \prod_{j=0}^{m} (\zeta_k - \omega_j) \quad \text{and} \quad \upsilon_k = \sum_{j=0}^{m} (\zeta_k - \omega_j)^{-2} .$$

The determinant $D$ may now be transformed into

$$(23): \qquad D = e^{-(\omega_1 + \omega_2 + \ldots + \omega_m)} \left| R_{10}\binom{\Omega}{\mathrm{P}_h}, R_{21}\binom{\Omega}{\mathrm{P}_h}, \ldots, R_{m,m-1}\binom{\Omega}{\mathrm{P}_h}, A_m\binom{\Omega}{\mathrm{P}_h} \right|_{h=0,1,\ldots,m} .$$

To do so, multiply for each suffix $k = 1, 2, \ldots, m$ the $(k+1)$st column of $D$ by the factor $e^{\omega_{k-1} - \omega_k}$ and subtract the result from the $k$th column, using the basic identities

$$A_{k-1}\binom{\Omega}{\mathrm{P}_h} - e^{\omega_{k-1} - \omega_k} A_k\binom{\Omega}{\mathrm{P}_h} = e^{-\omega_k} R_{k,k-1}\binom{\Omega}{\mathrm{P}_h} \qquad \binom{h = 0,1,\ldots,m}{k = 1,2,\ldots,m} .$$

**14.** On substituting in (23) for $A_m$ and the $R_{k,k-1}$ their asymptotic expressions, it follows that

$$D = \left[\left(\frac{2\pi}{\rho}\right)^m (\upsilon_1\upsilon_2\ldots\upsilon_m)^{-1}\right]^{1/2} (\tau_1\tau_2\ldots\tau_m)^{\rho} \big((m+1)\rho\big)! \times$$

$$\times \left\{ \left| z_{h1}, z_{h2}, \ldots, z_{hm}, \big((m+1)\rho\big)^{\sum_{j=0}^{m} \theta_{hj}} \right|_{j=0,1,\ldots,m} + o(1) \right\} .$$

Here $o(1)$ tends to zero as $\rho$ tends to infinity.

Choose in particular

$$\theta_{hk} = 0 \quad \text{for} \quad k = 0, 1, \ldots, m-1 \ ; \quad \theta_{hm} = h \qquad (h = 0,1,\ldots,m)$$

so that $\sum_{j=0}^{m} \theta_{hj} = h$ assumes its largest value when $h = m$. Since $\rho$ tends to infinity, the preceding determinant for $D$ can therefore be simplified, and we find that

$$D = \left\{\left(\frac{2\pi}{\rho}\right)^m (\upsilon_1\upsilon_2\ldots\upsilon_m)^{-1}\right\}^{1/2} (\tau_1\tau_2\ldots\tau_m)^\rho \left((m+1)\rho\right)! \left((m+1)\rho\right)^m \times$$
$$\times \left\{\left|(\zeta_1-\omega_m)^h, (\zeta_2-\omega_m)^h, \ldots, (\zeta_m-\omega_m)^h\right|_{h=0,1,\ldots,m-1} + o(1)\right\} .$$

Here the Vandermonde determinant on the right-hand side is identical with the difference product of the numbers $\zeta_1-\omega_m, \zeta_2-\omega_m, \ldots, \zeta_m-\omega_m$ and so is distinct from $0$ since, by construction, all the roots $\zeta_1, \zeta_2, \ldots, \zeta_m$ are distinct. Hence, as soon as $\rho$ is sufficiently large, $D$ is distinct from zero, and it follows that at least one of the sums

$$S(\rho,\ldots,\rho,\rho+h) \qquad (h = 0,1,\ldots,m)$$

likewise is not zero. In the special case when the $\omega_k$ are distinct rational integers and the $c_k$ are rational non-vanishing integers, this implies by Theorem 1 the transcendency of $e$.

Strangely enough, Hermite himself took slightly different values for the $\theta_{hk}$, namely

$$\theta_{hk} = h \qquad (h,k = 0,1,\ldots,m) .$$

Now one obtains for $D$ the formula

$$D = \left\{\left(\frac{2\pi}{\rho}\right)^m (\upsilon_1\upsilon_2\ldots\upsilon_m)^{-1}\right\}^{1/2} (\tau_1\tau_2\ldots\tau_m)^\rho \left((m+1)\rho\right)! \left((m+1)\rho\right)^{m(m+1)} \times$$
$$\times \left\{\left|\tau_1^h, \tau_2^h, \ldots, \tau_m^h\right|_{h=0,1,\ldots,m-1} + o(1)\right\} .$$

Here the new determinant is the difference product of the numbers $\tau_1, \tau_2, \ldots, \tau_m$. However, these numbers need not always be all distinct, and therefore *with this choice of the integers* $\theta_{hk}$ *it remains in doubt whether* $D$ *does or does not vanish for sufficiently large values of* $\rho$.

**15.** On account of this difficulty, Hermite gave a second proof for the non-vanishing of $D$, choosing, with Kronecker's symbol $\delta_{hk}$,

$$\rho_{hk} = \rho - \delta_{hk} \qquad (h,k = 0,1,\ldots,m) .$$

The proof is very elegant and of interest in itself, but slightly complicated. It leads to the explicit value of $D$ for this choice of the $\rho$'s .

Hermite first constructs a polynomial

$$\Theta_\rho(y,z) = \Theta_\rho(y,z|\omega_0,\omega_1,\ldots,\omega_m)$$

in $y$, $z$, $\rho$, $\omega_0$, $\omega_1$, ..., $\omega_m$ with the following properties.

(a): *As a function of* $y$ *and* $z$ *,* $\Theta_\rho$ *has the form*

$$\Theta_\rho(y,z) = y^m + y^{m-1}z + y^{m-2}z^2 + \ldots + z^m + \Theta_\rho^*(y,z)$$

*where* $\Theta_\rho^*$ *is a polynomial in* $y$ *and* $z$ *which does not contain terms of higher dimension than* $m-1$ *in* $y$ *and* $z$ .

(b): *For* $h = 0, 1, \ldots, m$ *, identically*

$$\int \prod_{l=0}^{m} (y-\omega_l)^{\rho-\delta_{hl}+1} e^{-y}dy - \rho \sum_{j=0}^{m} \Theta_\rho(\omega_h,\omega_j) \int \prod_{l=0}^{m} (y-\omega_l)^{\rho-\delta_{jl}} e^{-y}dy =$$

$$= -e^{-y}.\Theta_\rho(\omega_h,y) \prod_{l=0}^{m} (y-\omega_l)^\rho .$$

Here, for each suffix $k = 0, 1, \ldots, m$ , integrate over a suitable path between $\omega_k$ and $\infty$ such that the right-hand side becomes $0$ . By the integral definition of $A_k$ the identities (b) imply then the set of $(m+1)^2$ equations

$$(24): \quad A_k\begin{pmatrix} \omega_0, & \omega_1, & \ldots, & \omega_m \\ \rho-\delta_{h0}+1, & \rho-\delta_{h1}+1, & \ldots, & \rho-\delta_{hm}+1 \end{pmatrix} =$$

$$= \rho \sum_{j=0}^{m} \Theta_\rho(\omega_h,\omega_j) A_k\begin{pmatrix} \omega_0, & \omega_1, & \ldots, & \omega_m \\ \rho-\delta_{j0}, & \rho-\delta_{j1}, & \ldots, & \rho-\delta_{jm} \end{pmatrix} \qquad (h,k = 0,1,\ldots,m) .$$

Introduce here the two square matrices of order $m+1$ ,

$$A\binom{\Omega}{\rho} = \left(A_k\begin{pmatrix} \omega_0, & \omega_1, & \ldots, & \omega_m \\ \rho-\delta_{h0}, & \rho-\delta_{h1}, & \ldots, & \rho-\delta_{hm} \end{pmatrix}\right)_{h,k=0,1,\ldots,m}$$

and

$$\Theta\binom{\Omega}{\rho} = (\Theta_\rho(\omega_h,\omega_k))_{h,k=0,1,\ldots,m} .$$

The set of equations (24) is then equivalent to the matrix equation

$$A\binom{\Omega}{\rho+1} = \rho\Theta\binom{\Omega}{\rho} A\binom{\Omega}{\rho} ,$$

and on applying this formula repeatedly, it follows that more generally

(25): $$A\binom{\Omega}{\rho} = (\rho-1)!\Theta\binom{\Omega}{\rho-1}\Theta\binom{\Omega}{\rho-2} \cdots \Theta\binom{\Omega}{1}A\binom{\Omega}{1} .$$

The polynomial $\Theta_\rho(y,z)$ retains its meaning for $\rho = 0$, but then the relations (b) reduce to

$$\int \prod_{l=0}^{m} (y-\omega_l)^{1-\delta_h} e^{-y}dy = -e^{-y}.\Theta_0(\omega_h,y) \qquad (h = 0,1,\ldots,m) ,$$

whence, for $k = 0, 1, \ldots, m$ ,

$$A_k\begin{bmatrix}\omega_0, & \omega_1, & \ldots, & \omega_m \\ 1-\delta_{h0}, & 1-\delta_{h1}, & \ldots, & 1-\delta_{hm}\end{bmatrix} = \Theta_0(\omega_h,\omega_k) \text{ and therefore } A\binom{\Omega}{0} = \Theta\binom{\Omega}{0} .$$

The equation (25) can then be replaced by

$$A\binom{\Omega}{\rho} = (\rho-1)!\Theta\binom{\Omega}{\rho-1}\Theta\binom{\Omega}{\rho-2} \cdots \Theta\binom{\Omega}{1}\Theta\binom{\Omega}{0} .$$

In this matrix formula take on both sides the determinants. If $D\binom{\Omega}{\rho}$ and $T\binom{\Omega}{\rho}$ denote the determinants of $A\binom{\Omega}{\rho}$ and $\Theta\binom{\Omega}{\rho}$, respectively, it follows that

$$D\binom{\Omega}{\rho} = \big((\rho-1)!\big)^{m+1}T\binom{\Omega}{\rho-1}T\binom{\Omega}{\rho-2} \cdots T\binom{\Omega}{1}T\binom{\Omega}{0} .$$

But, by the property (a) of $\Theta_\rho(y,z)$, it follows easily that, independent of $\rho$, the determinant $T\binom{\Omega}{\rho}$ has the value

$$T\binom{\Omega}{\rho} = \prod_{\substack{h,k=0 \\ h\neq k}}^{m} (\omega_k-\omega_h) .$$

Therefore finally,

(26): $$D\binom{\Omega}{\rho} = \big((\rho-1)!\big)^{m+1} \prod_{\substack{h,k=0 \\ h\neq k}}^{m} (\omega_k-\omega_h)^\rho .$$

This elegant formula by Hermite shows at once that $D\binom{\Omega}{\rho}$ does not vanish if the numbers $\omega_0, \omega_1, \ldots, \omega_m$ are all distinct. Hence, if $o_0, o_1, \ldots, o_m$ are distinct from zero, then to every positive integer $\rho$ there exists at least one suffix $h$ for which the sum $S_h(\rho)$ does not vanish.

For rational integral $\omega_k$ and $o_k$ this implies again the transcendency of $e$. If, however, the parameters $\omega_k$ and $o_k$ satisfy the Hypothesis 3, then, by Theorem 3, this result leads at once to the General Theorem of Lindemann and to the transcendency of $\pi$. This was essentially Lindemann's method.

**16.** There is a much simpler way of proving Hermite's formula (26) and in fact a more general formula. This was seemingly first shown by K. Mahler in 1931 (see also his paper of 1968 where the problem is placed in a wider context).

For Mahler's proof, the single positive integer $\rho$ may be replaced by an arbitrary vector

$$P = (\rho_0, \rho_1, \ldots, \rho_m)$$

with positive integral components $\rho_k$, and the variable $x$ may be retained. Thus for $h, k, l = 0, 1, \ldots, m$, let

$$A_{hk}\left(x\middle|{\Omega \atop P}\right) = A_k\left(x\middle|{\omega_0, \quad \omega_1, \quad \ldots \quad \omega_m \atop \rho_0-\delta_{h0}, \rho_1-\delta_{h1}, \ldots, \rho_m-\delta_{hm}}\right)$$

and

$$R_{h,kl}\left(x\middle|{\Omega \atop P}\right) = R_{kl}\left(x\middle|{\omega_0, \quad \omega_1, \quad \ldots, \quad \omega_m \atop \rho_0-\delta_{h0}, \rho_1-\delta_{h1}, \ldots, \rho_m-\delta_{hm}}\right).$$

Then for all suffices $h, k, l$,

$$A_{kl}\left(x\middle|{\Omega \atop P}\right)e^{\omega_k x} - A_{hk}\left(x\middle|{\Omega \atop P}\right)e^{\omega_l x} = R_{h,kl}\left(x\middle|{\Omega \atop P}\right).$$

Further, denote by $D\left(x\middle|{\Omega \atop P}\right)$ the determinant of order $m+1$,

$$D\left(x\middle|{\Omega \atop P}\right) = \left|A_{hk}\left(x\middle|{\Omega \atop P}\right)\right|_{h,k=0,1,\ldots,m}.$$

This determinant can be evaluated by means of the properties (8) of §5 and (11) of §6, as follows.

By (8), its diagonal elements have the form

$$A_{kk}\left(x\middle|{\Omega \atop P}\right) = (\rho_k-1)!\,x^{\sigma-\rho_k}\prod_{\substack{h=0\\h\neq k}}^{m}(\omega_k-\omega_h)^{\rho_h} \text{ plus terms in lower powers of } x;$$

here $\sigma = \rho_0 + \rho_1 + \ldots + \rho_m$. On the other hand, $A_{hk}\left(x\middle|{\Omega \atop P}\right)$ is at most of degree $\sigma - \rho_k - 1$ if $h \neq k$. Hence $D\left(x\middle|{\Omega \atop P}\right)$ is a polynomial in $x$ of the form

$$(27):\quad D\left(x\middle|{\Omega \atop P}\right) = \left\{\sum_{k=0}^{m}(\rho_k-1)!\right\}x^{m\sigma}\prod_{\substack{h,k=0\\h\neq k}}^{m}(\omega_k-\omega_h)^{\rho_h} \text{ plus terms in lower powers of } x.$$

On the other hand, by (11), all the expressions $R_{h,kl}\left(x\middle|{\Omega \atop P}\right)$ are entire functions of $x$ which have at the origin $x = 0$ a zero at least of order $\sigma$.

Furthermore a transformation analogous to that in §13 leads to the formula

$$D\left(x\middle|\begin{smallmatrix}\Omega\\ \rho\end{smallmatrix}\right) = e^{-(\sigma-\rho_0)x}\left|R_{h,10}\left(x\middle|\begin{smallmatrix}\Omega\\ \rho\end{smallmatrix}\right), R_{h,21}\left(x\middle|\begin{smallmatrix}\Omega\\ \rho\end{smallmatrix}\right), \ldots, R_{h,m,m-1}\left(x\middle|\begin{smallmatrix}\Omega\\ \rho\end{smallmatrix}\right), A_{hm}\left(x\middle|\begin{smallmatrix}\Omega\\ \rho\end{smallmatrix}\right)\right|_{h=0,1,\ldots,m},$$

which shows immediately that $D\left(x\middle|\begin{smallmatrix}\Omega\\ \rho\end{smallmatrix}\right)$ has at $x = 0$ a zero at least of order $m\sigma$. Hence finally by (27),

$$(28): \qquad D\left(x\middle|\begin{smallmatrix}\Omega\\ \rho\end{smallmatrix}\right) = \left\{\prod_{k=0}^{m}(\rho_k-1)!\right\}x^{m\sigma}\prod_{\substack{h,k=0\\ h\neq k}}^{m}(\omega_k-\omega_h)^{\rho_h}.$$

In the special case when $x = 1$ and $\rho_0 = \rho_1 = \ldots = \rho_m = \rho$ this becomes Hermite's formula (26).

**17.** While Hermite established asymptotic formulae and even explicit expressions for certain determinants $D$, O. Venske (1890) and K. Weierstrass (1885) were satisfied with proving that certain determinants of this type are distinct from zero.

Venske assumes again, as in Hermite's first proof (§15), that $\omega_0, \omega_1, \ldots, \omega_m$ are real and distinct, hence that without loss of generality

$$\omega_0 < \omega_1 < \ldots < \omega_m .$$

He then proves the following result.

(29): *Let* $r_{hk}$, *for* $h, k = 0, 1, \ldots, m$, *be a system of* $(m+1)^2$ *arbitrary positive integers. Then one can select* $(m+1)^2$ *integers* $e_{hk}$ *with the values* 0 *or* 1 *such that the determinant* $D$ *belonging to the parameters*

$$\rho_{hk} = r_{hk} + e_{hk} \qquad (h,k = 0,1,\ldots,m)$$

*does not vanish.*

In the proof of this result, let as usual

$$\operatorname{sgn} x = \begin{cases} +1 & \text{for } x > 0, \\ 0 & \text{for } x = 0, \\ -1 & \text{for } x < 0. \end{cases}$$

Further put

$$J_{hk} = \begin{cases} R_{k+1,k}\begin{pmatrix}\omega_0, \omega_1, \ldots, \omega_m \\ \rho_{h0}, \rho_{h1}, \ldots, \rho_{hm}\end{pmatrix} & \text{if } h = 0, 1, \ldots, m \; ; \; k = 0, 1, \ldots, m-1 \; ; \\ \\ A_m\begin{pmatrix}\omega_0, \omega_1, \ldots, \omega_m \\ \rho_{h0}, \rho_{h1}, \ldots, \rho_{hm}\end{pmatrix} & \text{if } h = 0, 1, \ldots, m \; ; \; k = m \; . \end{cases}$$

The integral formulae

$$R_{k+1,k}\begin{pmatrix}\omega_0, \omega_1, \ldots, \omega_m \\ \rho_{h0}, \rho_{h1}, \ldots, \rho_{hm}\end{pmatrix} = e^{\omega_{k+1}+\omega_k} \int_{\omega_k}^{\omega_{k+1}} \prod_{l=0}^{m} (y-\omega_l)^{\rho_{hl}} e^{-y} dy$$

and

$$A_m\begin{pmatrix}\omega_0, \omega_1, \ldots, \omega_m \\ \rho_{h0}, \rho_{h1}, \ldots, \rho_{hm}\end{pmatrix} = e^{\omega_m} \int_{\omega_m}^{\infty} \prod_{l=0}^{m} (y-\omega_l)^{\rho_{hl}} e^{-y} dy$$

evidently imply the signs

$$\operatorname{sgn} J_{hk} = \begin{cases} (-1)^{\rho_{h,k+1}+\rho_{h,k+2}+\cdots+\rho_{hm}} & \text{if } h = 0, 1, \ldots, m \; ; \; k = 0, 1, \ldots, m-1 \; ; \\ \\ +1 & \text{if } h = 0, 1, \ldots, m \; ; \; k = m \; . \end{cases}$$

Now let $n$ be any one of the suffixes $0, 1, \ldots, m$ , and let $\Delta_n$ be the determinant

$$\Delta_n = \left| J_{hk} \right|_{h,k=n,n+1,\ldots,m} .$$

In the same way as in §13 and §16, one can show that the determinant

$$D = \left| A_k\begin{pmatrix}\omega_0, \omega_1, \ldots, \omega_m \\ \rho_{h0}, \rho_{h1}, \ldots, \rho_{hm}\end{pmatrix} \right|_{h,k=0,1,\ldots,m}$$

is connected with $\Delta_0$ by the identity

$$\Delta_0 = e^{\omega_1+\omega_2+\cdots+\omega_m} D .$$

Hence, in particular,

$$\operatorname{sgn} \Delta_0 = \operatorname{sgn} D .$$

It is further obvious from $\Delta_m = J_{mm} > 0$ that

$$\operatorname{sgn} \Delta_m = 1 .$$

Now let $\Delta_{nk}$ , for $k = n, n+1, \ldots, m$ , denote the cofactor of the element $J_{nk}$

in the first row of the determinant $\Delta_n$ . Then evidently

$$\Delta_{nn} = \Delta_{n+1}$$

and

(30): $$\Delta_n = \sum_{k=n}^{m} J_{nk}\Delta_{nk} \ .$$

Here $\Delta_n$ depends only on the parameters $\rho_{hk}$ for which $h = n, n+1, \ldots, m$ ; and each of the cofactors $\Delta_{nn}, \Delta_{n,n+1}, \ldots, \Delta_{nm}$ depends only on the parameters $\rho_{hk}$ for which $h = n+1, n+2, \ldots, m$ .

Choose

$$e_{hk} = 0 \quad \text{if} \quad h \geq k \ .$$

Venske bases his proof on the following recursive result.

*If, for any suffix* $n = 0, 1, \ldots, m-1$ , *the integers* $e_{hk}$ *with* $h \geq n+1$ *have already been fixed such that* $\Delta_{n+1} \neq 0$ , *then it is possible to select*

$$e_{n,n+1}, e_{n,n+2}, \ldots, e_{nm}$$

*equal to* 0 *or* 1 *such that also* $\Delta_n \neq 0$ .

For by the value of $\operatorname{sgn} J_{hk}$ ,

$$\operatorname{sgn}(J_{nk}\Delta_{nk}) = (-1)^{\sum_{l=k+1}^{m}(r_{nl}+e_{nl})} \operatorname{sgn} \Delta_{nk} \ .$$

Here, by hypothesis, $\Delta_{nn} = \Delta_{n+1}$ does not vanish and therefore also $J_{nn}\Delta_{nn} \neq 0$ . We can now successively give values 0 or 1 to $e_{nm}, e_{n,m-1}, \ldots, e_{n,n+1}$ such that all those products $J_{nk}\Delta_{nk}$ in (30) for which $\Delta_{nk} \neq 0$ have the *same sign* as $J_{nn}\Delta_{nn}$ . From this Venske's recursive result follows at once.

Since $\Delta_m \neq 0$ , by successively applying this result for $n = m-1, m-2, \ldots, 0$ , it follows that, if the values 0 or 1 of the $e_{hk}$ are chosen suitably, all the determinants $\Delta_{m-1}, \Delta_{m-2}, \ldots, \Delta_0$ can be made distinct from zero, and so also $D \neq 0$ .

18. Up to now, the integer $p$ was equal to $m$ . Weierstrass (1885) assumes that $m = p + 1$ and proves the following result.

(31): *Let* $\omega_0, \omega_1, \ldots, \omega_{p+1}$ *be any complex numbers such that* $\omega_0, \omega_1, \ldots, \omega_p$ *are all distinct; let* $c_0, c_1, \ldots, c_p$ *be arbitrary complex numbers not all zero, and let* $\rho$ *be any positive integer. Then at least one of the* $p+1$ *sums*

$$S(\rho;h) = \sum_{k=0}^{p} A_k\begin{pmatrix}\omega_0, \omega_1, \ldots, \omega_p, \omega_{p+1}\\ \rho, \rho, \ldots, \rho \quad h\end{pmatrix} c_k \qquad (h = 0,1,\ldots,p)$$

*is distinct from zero.*

The proof depends on simple divisibility properties of polynomials. Denote by

$$C_0, C_1, \ldots, C_p$$

any $p+1$ complex parameters which do not all vanish. Further put

$$f(y) = \prod_{k=0}^{p} (y-\omega_k), \quad F(y) = f(y)^{\rho} \sum_{h=0}^{p} C_h (y-\omega_{p+1})^h, \quad F[y] = \sum_{\lambda=0}^{\infty} \frac{d^{\lambda}F(y)}{dy^{\lambda}}$$

so that evidently

$$F[\omega_k] = \sum_{h=0}^{p} C_h A_k\begin{pmatrix}\omega_0, \omega_1, \ldots, \omega_p, \omega_{p+1}\\ \rho, \rho, \ldots, \rho, \quad h\end{pmatrix} \qquad (k = 0,1,\ldots,p) .$$

The expression $F[y]$ is a polynomial in $y$ of degree at most

$$(p+1)\rho + p ,$$

and it is obvious from its definition that it satisfies the differential equation

(32): $$F[y] - \frac{dF[y]}{dy} = F(y) .$$

Assume that $f(y)^{\kappa}$ is the highest integral power of $f(y)$ which divides $F[y]$ ; thus

$$0 \leq \kappa \leq \rho .$$

*We assert that in fact* $\kappa = 0$ . For assume that $\kappa \geq 1$ . All the zeros of $f(y)$ by hypothesis are distinct; hence $f(y)^{\kappa-1}$ is the highest power of $f(y)$ which divides $\frac{dF[y]}{dy}$ and hence also is the highest power of $f(y)$ which divides the left-hand side of (32). Since the right-hand side is divisible by $f(y)^{\rho}$ , this is impossible.

Since then $F[y]$ is not divisible by $f(y)$ , at least one of the numbers

$$F[\omega_k] \qquad (k = 0,1,\ldots,p)$$

is distinct from zero whenever the coefficients $C_0, C_1, \ldots, C_p$ do not vanish simultaneously. But this implies that the determinant

$$\left| A_k \begin{pmatrix} \omega_0, \omega_1, \ldots, \omega_p, \omega_{p+1} \\ \rho, \rho, \ldots, \rho, h \end{pmatrix} \right|_{h,k=0,1,\ldots,p}$$

does not vanish and that therefore *at least one of the sums*

$$S(\rho;h) \qquad\qquad (h = 0,1,\ldots,p)$$

*is distinct from zero.*

On specialising the parameters and using the theorems of section II, we derive again the theorems of Hermite and Lindemann.

This proof by Weierstrass of the property (31) is particularly elegant. Weierstrass's property has also been used in proofs of the transcendency of $e$ and $\pi$ by F. Mertens (1896) and F. Schottky (1914).

**19.** In the proofs of the non-vanishing of suitably chosen determinants $D$ as given in this section, both analytic and algebraic methods have been applied. Instead one might make use of arithmetic considerations. This has in fact been done by H. Weber (1899). His method will be described in the next section.

## IV

## Proofs depending on Divisibility Properties

**20.** D. Hilbert (1893) was the first to use arithmetical properties of the polynomials $A_k\left(x \middle| \begin{smallmatrix}\Omega\\ \mathrm{P}\end{smallmatrix}\right)$ to prove the transcendency of $e$ and $\pi$.

Again put $x = 1$. The property (10) of §5 may be stated as follows.

**(10'):** $A_k\begin{pmatrix}\Omega\\ \mathrm{P}\end{pmatrix} = A_k\begin{pmatrix}\omega_0, \omega_1, \ldots, \omega_m \\ \rho_0, \rho_1, \ldots, \rho_m\end{pmatrix}$ *is a polynomial in* $\omega_0, \omega_1, \ldots, \omega_m$ *with rational integral coefficients divisibly by* $\rho_k!$. *Furthermore,*

$$A_k\begin{pmatrix}\Omega\\ \mathrm{P}\end{pmatrix} = \rho_k! \prod_{\substack{h=0\\ h\neq k}}^{m} (\omega_k - \omega_h)^{\rho_h} + (\rho_k+1)! A_k^*\begin{pmatrix}\Omega\\ \mathrm{P}\end{pmatrix},$$

*where also* $A_k^*\begin{pmatrix}\Omega\\ \mathrm{P}\end{pmatrix} = A_k^*\begin{pmatrix}\omega_0, \omega_1, \ldots, \omega_m \\ \rho_0, \rho_1, \ldots, \rho_m\end{pmatrix}$ *is a polynomial in* $\omega_0, \omega_1, \ldots, \omega_m$ *with rational integral coefficients.*

We further note that, by the property (9) of §5,

**(9'):** $A_k\begin{pmatrix}\Omega\\ \mathrm{P}\end{pmatrix}$ *and* $A_k^*\begin{pmatrix}\Omega\\ \mathrm{P}\end{pmatrix}$ *contain only terms at most of dimension*

$$\sum_{h=0}^{m} \rho_h - \rho_k < \sum_{h=0}^{m} \rho_h + 1$$

*in* $\omega_0, \omega_1, \ldots, \omega_m$ .

Now let $\omega_0, \omega_1, \ldots, \omega_m$ be algebraic numbers, and let $N$ be a positive integer such that the products $N\omega_0, N\omega_1, \ldots, N\omega_m$ are algebraic integers. On putting again

$$\sigma = \sum_{h=0}^{m} \rho_h ,$$

both

$$N^{\sigma+1} A_k \binom{\Omega}{\rho} \quad \text{and} \quad N^{\sigma+1} A_k^* \binom{\Omega}{\rho}$$

are algebraic integers, and the first of these integers satisfies the congruence

(33): $$N^{\sigma+1} A_k \binom{\Omega}{\rho} \equiv N^{\rho_k+1} \rho_k! \prod_{\substack{h=0 \\ h\neq k}}^{m} (N\omega_k - N\omega_h)^{\rho_h} \pmod{(\rho_k+1)!} .$$

This means, in particular, that the left-hand side is an algebraic integer divisible by $\rho_k!$ .

21. As a first application, let $\omega_0, \omega_1, \ldots, \omega_m$ and $o_0, o_1, \ldots, o_m$ be *rational integers* with the properties (a) and (b) of Hypothesis 1; here we have put $p = m$ . If the positive integers $\rho_0, \rho_1, \ldots, \rho_m$ are defined by

$$\rho_0 = \rho - 1 ; \quad \rho_k = \rho \quad \text{for} \quad k = 1, 2, \ldots, m ,$$

where $\rho$ denotes a sufficiently large positive integer, then also the property (c) of Hypothesis 1 is satisfied. Hence, by Theorem 1, $e$ is proved to be transcendental if there exist arbitrarily large values of $\rho$ for which the sum

$$S(\rho-1,\rho,\ldots,\rho) = \sum_{k=0}^{m} A_k \begin{pmatrix} \omega_0, \omega_1, \ldots, \omega_m \\ \rho-1, \rho, \ldots, \rho \end{pmatrix} o_k$$

does not vanish. In the present case we may choose $N = 1$ , and by (33) we obtain the congruences

$$A_0 \begin{pmatrix} \omega_0, \omega_1, \ldots, \omega_m \\ \rho-1, \rho, \ldots, \rho \end{pmatrix} \equiv (\rho-1)! \prod_{h=1}^{m} (\omega_0 - \omega_h)^{\rho} \pmod{\rho!} ,$$

but

$$A_k\begin{pmatrix}\omega_0,\omega_1,\ldots,\omega_m\\ \rho-1,\ \rho,\ldots,\ \rho\end{pmatrix} \equiv 0 \pmod{\rho!} \qquad (k = 1,2,\ldots,m) .$$

These congruences imply that

$$S(\rho-1,\rho,\ldots,\rho) \equiv c_0(\rho-1)! \prod_{h=1}^{m} (\omega_0-\omega_h)^\rho \pmod{\rho!} .$$

Now take for $\rho$ a sufficiently large positive integer which is relatively prime to the $m+1$ integers

$$c_0,\ \omega_0-\omega_1,\ \omega_0-\omega_2,\ \ldots,\ \omega_0-\omega_m .$$

Then

$$\frac{1}{(\rho-1)!} S(\rho-1,\rho,\ldots,\rho)$$

is an integer not divisible by $\rho$, hence cannot be $0$. The transcendency of $e$ follows then immediately.

22. A somewhat similar proof is given by Hilbert for the transcendency of $\pi$. It is based on Euler's equation

$$e^{\pi i} + 1 = 0$$

which shows that it suffices to prove the following result.

*Let* $\alpha$ *be any algebraic number, and let* $\alpha^{(1)} = \alpha,\ \alpha^{(2)},\ \ldots,\ \alpha^{(n)}$ *for all its algebraic conjugates. Then the product*

$$P = \prod_{h=1}^{n} \left(e^{\alpha^{(h)}}+1\right) = 1 + \sum_{\substack{0\le i_1<i_2<\ldots<i_j\le n\\ 1\le j\le n}} e^{\alpha^{(i_1)}+\alpha^{(i_2)}+\ldots+\alpha^{(i_j)}}$$

*does not vanish.*

For $P$ can be written as

$$P = c_0 + e^{\omega_1} + e^{\omega_2} + \ldots + e^{\omega_m} .$$

Here $c_0$ is equal to the term $1$ plus the number of those sums

$$\alpha^{(i_1)} + \alpha^{(i_2)} + \ldots + \alpha^{(i_j)}$$

which are equal to zero, while $\omega_1,\ \omega_2,\ \ldots,\ \omega_m$ denote those of these sums in some order which do not vanish. It is thus evident that $c_0$ is a positive integer, and that $\omega_1,\ \omega_2,\ \ldots,\ \omega_m$ are algebraic numbers not necessarily all distinct. From the definition of these numbers it is clear that they can be split into finitely many sets $\Sigma_1, \Sigma_2, \ldots, \Sigma_u$ where, as in §9, each of these sets consists of exactly one algebraic number and all its algebraic conjugates; however, in the present case, these sets need not necessarily all be distinct. It follows that

*Every symmetric function of* $\omega_1, \omega_2, \ldots, \omega_m$ *with rational coefficients is itself a rational number.*

Thus, in particular, if $\rho$ is any positive integer, then both expressions

$$(34): \qquad A_0\begin{pmatrix}0, \omega_1,\ldots,\omega_m\\ \rho-1, \rho,\ldots, \rho\end{pmatrix}c_0 \quad\text{and}\quad \sum_{k=1}^{m} A_k\begin{pmatrix}0, \omega_1,\ldots,\omega_m\\ \rho-1, \rho,\ldots, \rho\end{pmatrix}$$

are rational numbers and hence also their sum

$$S(\rho-1,\rho,\ldots,\rho) = A_k\begin{pmatrix}0, \omega_1,\ldots,\omega_m\\ \rho-1, \rho,\ldots, \rho\end{pmatrix}c_0 + \sum_{k=1}^{m} A_k\begin{pmatrix}0, \omega_1,\ldots,\omega_m\\ \rho-1, \rho,\ldots, \rho\end{pmatrix} .$$

Let $N$ be a positive integer such that $N\omega_1, \ldots, N\omega_m$ are algebraic integers, so that the product of norms $\prod_{h=1}^{m}(-N\omega_h)$ is a rational integer distinct from zero. Furthermore, the multiples of the two numbers (34) by $N^{(m+1)\rho}$ are rational algebraic integers and so are rational integers. These integers, by (33), satisfy the congruences

$$N^{(m+1)\rho}A_0\begin{pmatrix}0, \omega_1,\ldots,\omega_m\\ \rho-1, \rho,\ldots, \rho\end{pmatrix}c_0 \equiv N^{\rho}(\rho-1)!c_0\prod_{h=1}^{m}(-N\omega_h)^{\rho} \pmod{\rho!}$$

and

$$\sum_{k=1}^{m} N^{(m+1)\rho}A_k\begin{pmatrix}0, \omega_1,\ldots,\omega_m\\ \rho-1, \rho,\ldots, \rho\end{pmatrix} \equiv 0 \pmod{\rho!} ,$$

so that also

$$N^{(m+1)\rho}S(\rho-1,\rho,\ldots,\rho) \equiv N^{\rho}(\rho-1)!c_0\prod_{h=1}^{m}(-N\omega_h)^{\rho} \pmod{\rho!} .$$

Finally, choose for $\rho$ a sufficiently large positive integer which is relatively prime to the rational integers

$$c_0,\ N,\ \text{and}\ \prod_{h=1}^{m}(-N\omega_h) .$$

Then

$$\frac{N^{(m+1)\rho}S(\rho-1,\rho,\ldots,\rho)}{(\rho-1)!}$$

is a rational integer which is not divisible by $\rho$ and so is distinct from zero. It follows that $\pi$ is transcendental.

**23.** It was essential in the proof just given that we were dealing with an equation

$$o_0 e^{\omega_0} + o_1 e^{\omega_1} + \ldots + o_p e^{\omega_p} = 0 \qquad (1 \le p \le m) ,$$

in which the first exponent $\omega_0$ is a *rational* number, while otherwise the conditions of Hypotheses 2 or 3 are satisfied. If one wants to prove the General Theorem of Lindemann by Hilbert's method, it would seem to be necessary first to transform the given equation into another in which $\omega_0$ has this property of being rational, in addition to the other properties of either of the two hypotheses.

However, a slight change of method obviates such a transformation. All that is necessary is to consider, not a single sum $S$, but instead a system of such sums of the form

$$S(\rho_{h0},\rho_{h1},\ldots,\rho_{hm}) = \sum_{k=0}^{p} A_k\begin{pmatrix}\omega_0, \omega_1,\ldots, \omega_m\\ \rho_{h0},\rho_{h1},\ldots,\rho_{hm}\end{pmatrix} o_k \qquad (h = 0,1,\ldots,p) .$$

If a suitable integral multiple of at least one of these sums can be proved not to be divisible by a given integer, it follows again that this sum does not vanish.

**24.** By way of example, assume that $p = m$ and that the parameters $\omega_k$, $o_k$, and $\rho$ satisfy the conditions of Hypothesis 3; let $N$ be a positive integer such that all the products $N\omega_0$, $N\omega_1$, ..., $N\omega_m$ are algebraic integers, and let $\rho$ assume sufficiently large values so that the condition (c) holds. By imposing a simple additional condition on $\rho$, we can make certain that all the sums

$$S_h = S_h(\rho) = \sum_{k=0}^{m} A_k\begin{pmatrix}\omega_0, & \omega_1, & \ldots, \omega_m\\ \rho-\delta_{h0},\rho-\delta_{h1},\ldots,\rho-\delta_{hm}\end{pmatrix} o_k \qquad (h = 0,1,\ldots,m)$$

are distinct from zero.

For in the present case the products

$$N^{(m+1)\rho} A_k\begin{pmatrix}\omega_0, & \omega_1, & \ldots, & \omega_m\\ \rho-\delta_{h0},\rho-\delta_{h1},\ldots,\rho-\delta_{hm}\end{pmatrix} , \quad = a_{hk}\binom{\Omega}{\rho} \quad \text{say},$$

all are algebraic integers, and in the ring of integers of the algebraic number field over $\mathbb{Q}$ generated by the $\omega$'s , they satisfy the congruences

$$a_{hk}\binom{\Omega}{\rho} \equiv N^{\rho}(\rho-1)! \prod_{\substack{l=0\\ l\neq k}}^{m} (N\omega_k - N\omega_l)^{\rho} \pmod{\rho!} \quad \text{if} \quad h = k ,$$

but

$$a_{hk}\binom{\Omega}{\rho} \equiv 0 \pmod{\rho!} \quad \text{if} \quad h \neq k \;.$$

Hence

$$N^{(m+1)\rho}S_h(\rho) \equiv o_h N^\rho(\rho-1)! \prod_{\substack{l=0\\ l\neq h}}^{m} (N\omega_h - N\omega_l) \pmod{\rho!} \qquad (h = 0,1,\ldots,m) \;.$$

Now let $\rho$ be relatively prime to the finitely many integers

$$o_0,\ o_1,\ \ldots,\ o_m,\ N,\ N\omega_h - N\omega_k \qquad (h,k = 0,1,\ldots,m;\ h \neq k)$$

Then none of the algebraic integers

$$N^{(m+1)\rho}S_h(\rho)\left((\rho-1)!\right)^{-1}$$

is divisible by $\rho$ , and so all the $S_h(\rho)$ are distinct from zero.

**25.** The selection of the parameters $\rho_{hk}$ as in §24 does not seem to occur in the literature. H. Weber (1899) chooses these parameters as in Weierstrass's proof (§18). In a slightly simplified form, his proof runs as follows.

Let $m = p + 1$ and

$$\rho_{h0} = \rho_{h1} = \ldots = \rho_{hp} = \rho \;, \quad \rho_{h,p+1} = h \qquad (h = 0,1,\ldots,p) \;.$$

With this choice of the $\rho_{hk}$ , assume that $\omega_0,\ \omega_1,\ \ldots,\ \omega_{p+1},\ o_0,\ o_1,\ \ldots,\ o_p$ , and $\rho$ satisfy the conditions of the Hypothesis 2. The following considerations show that not all the sums

$$S(\rho,\ldots,\rho,h) = \sum_{k=0}^{p} A_k\begin{Bmatrix}\omega_0,\omega_1,\ldots,\omega_p,\omega_{p+1}\\ \rho,\ \rho,\ldots,\ \rho \quad h\end{Bmatrix} o_k \qquad (h = 0,1,\ldots,p)$$

are zero.

For $h,\ k = 0,\ 1,\ \ldots,\ p$ ,

$$N^{(p+1)\rho+h+1}A_k\begin{Bmatrix}\omega_0,\omega_1,\ldots,\omega_p,\omega_{p+1}\\ \rho,\ \rho,\ldots,\ \rho,\quad h\end{Bmatrix} \equiv N^{\rho+1}\rho! \sum_{\substack{l=0\\ l\neq k}}^{p} (N\omega_k - N\omega_l)^\rho (N\omega_k - N\omega_{p+1})^h \;\left(\mathrm{mod}\ (\rho+1)!\right) \;,$$

and therefore, for $h = 0,\ 1,\ \ldots,\ p$ ,

$$N^{(p+1)\rho+h+1}S(\rho,\ldots,\rho,h) \equiv N^{\rho+1}\rho! \sum_{k=0}^{p} g(N\omega_k)^\rho (N\omega_k - N\omega_{p+1})^h \;\left(\mathrm{mod}\ (\rho+1)!\right) \;,$$

with the same meaning of $N$ as before, and with $g(y)$ defined by

$$g(y) = \prod_{l=0}^{p} (y-N\omega_l) .$$

Conversely,

$$\Delta N^{\rho+1}\rho!g(N\omega_k)^{\rho}o_k \equiv N^{(p+1)\rho+1} \sum_{h=0}^{p} \Delta_{hk}N^h S(\rho,\dots,\rho,h) \pmod{(\rho+1)!} \qquad (k = 0,1,\dots,p) ,$$

where $\Delta$ denotes the Vandermonde determinant

$$\Delta = \left|(N\omega_k - N\omega_{p+1})^h\right|_{h,k=0,1,\dots,p} ,$$

and $\Delta_{hk}$ is the cofactor of its element $(N\omega_k-N\omega_{p+1})^h$ . Assume now that the positive integer $\rho + 1$ is relatively prime to all the algebraic integers

$$N,\ o_0,\ o_1,\ \dots,\ o_p,\ N\omega_h-N\omega_k \qquad (h,k = 0,1,\dots,p;\ h \neq k) .$$

Then none of the products

$$\Delta N^{\rho+1}g(N\omega_k)^{\rho}o_k \qquad (k = 0,1,\dots,p)$$

is divisible by $\rho + 1$ because $\Delta$ is the difference product of the numbers $N\omega_0$, $N\omega_1$, ..., $N\omega_p$ . Therefore at least one of the algebraic integers

$$S(\rho,\dots,\rho,h)(\rho!)^{-1} \qquad (h = 0,1,\dots,p)$$

also is not divisible by $\rho + 1$ and so cannot be equal to zero.

**26.** In all the proofs of this section, either $\rho$ or $\rho + 1$ was chosen sufficiently large and relatively prime to a finite number of algebraic integers. Denote by $M$ the product of the absolute values of the norms of these algebraic integers; thus $M$ is a positive integer. The condition for $\rho$ is now trivially satisfied if we take $\rho$ , or $\rho + 1$ , equal to

(35): $$Mr + 1$$

where $r$ is a sufficiently large positive integer.

A. Hurwitz (1893) satisfies the divisibility conditions in a seemingly simpler way by taking his parameter equal to a sufficietly large prime. This is a very convenient choice, but it imposes unnecessary restrictions on the parameter without actually simplifying the proof. The disadvantage of Hurwitz's choice becomes particularly evident if one wants to derive a measure of transcendency, say for $e$ . For instead of being able to select the parameter in the simple arithmetic progression (35), it now belongs to the highly complex sequence of primes, and one needs the prime number theorem in the estimates. See, for example, J. Popken (1929).

# V

## Proofs depending on Analytic Estimates

**27.** In the preceding proofs we always showed that a suitable sum $S$ was not zero. There are two proofs in the literature in which instead an inequality $T \neq 0$ is established. The proof by Th. Stieltjes (1890) leads only to the transcendency of $e$. It depends on fixing the sign of a certain integral and is similar in this respect to the considerations by Venske in §17. On the other hand, the method by H. Späth (1928) allows to prove the General Theorem of Lindemann. It depends on estimates for integrals which go back to Laplace's classical work on the asymptotic evaluation of integrals.

**28.** For the proof by Stieltjes, assume that $\omega_0, \omega_1, \ldots, \omega_m, o_0, o_1, \ldots, o_m$ satisfy the conditions (a) and (b) of Hypothesis 1 and that in addition

$$\omega_0 < \omega_1 < \ldots < \omega_m .$$

By definition

$$R_{k0}\begin{pmatrix}\omega_0,\omega_1,\ldots,\omega_m\\ \rho_0,\rho_1,\ldots,\rho_m\end{pmatrix} = e^{\omega_0+\omega_k}\int_{\omega_0}^{\omega_k} e^{-y}\prod_{l=0}^{m}(y-\omega_l)^{\rho_l}dy \qquad (k = 0,1,\ldots,m) .$$

Hence the sum

$$T(\rho_0,\rho_1,\ldots,\rho_m) = e^{-\omega_0}\sum_{k=0}^{m} R_{k0}\begin{pmatrix}\omega_0,\omega_1,\ldots,\omega_m\\ \rho_0,\rho_1,\ldots,\rho_m\end{pmatrix} o_k$$

can be written as

$$T(\rho_0,\rho_1,\ldots,\rho_m) = \int_{\omega_0}^{\omega_k} e^{-y}H(y)dy ,$$

where $H(y)$ denotes the evidently continuous function defined by

$$H(y) = \prod_{j=0}^{m}(y-\omega_j)^{\rho_j} \cdot \sum_{l=k}^{m} o_l e^{\omega_l} \quad \text{for} \quad \omega_{k-1} \leq y \leq \omega_k \qquad (k = 1,2,\ldots,m) .$$

Thus, in particular,

$$H(y) = \prod_{j=0}^{m}(y-\omega_j)^{\rho_j} . o_m e^{m} \quad \text{for} \quad \omega_{m-1} \leq y \leq \omega_m ,$$

so that since $o_m \neq 0$ the function $H(y)$ has for $\omega_{m-1} < y < \omega_m$ a fixed sign and does not vanish identically. Evidently

$$\operatorname{sgn} H(y) = (-1)^{\rho_k+\rho_{k+1}+\dots+\rho_m} \operatorname{sgn}\left(\sum_{l=k}^{m} c_l e^{\omega_l}\right) \quad \text{for } \omega_{k-1} < y < \omega_k \qquad (k = 1,2,\dots,m) .$$

Thus if the $\rho_k$ are restricted to suitable residue classes (mod 2) , then $H(y)$ on $\omega_0 \le y \le \omega_m$ is non-negative and not identically zero, and therefore $T(\rho_0,\rho_1,\dots,\rho_m)$ is positive. On assuming that $\rho_{\min}$ is sufficiently large, we obtain again the transcendency of $e$ .

29. Späth's method is based on the following lemma, which is a simple consequence of the first mean value theorem of integral calculus.

*Let $a < b$ , and let $\phi(y)$ and $\psi(y)$ be two continuous functions on the interval $a \le y \le b$ which satisfy the inequalities*

$$\phi(y) > 0 \quad \text{and} \quad \psi(y) \ge 0 .$$

*Assume that the maximum*

$$\mu = \max_{a \le y \le b} \psi(y)$$

*is positive, and denote by $\varepsilon$ a positive number less than $\mu$ , and by $n$ a sufficiently large positive integer. Then*

$$\int_a^b \phi(y)\psi(y)^n dy \ge (\mu-\varepsilon)^n .$$

We begin by applying this lemma to the transcendency of $e$ . Just as in Stieltjes's proof, let the $\omega_k$ and $c_k$ satisfy the conditions (a) and (b) of Hypothesis 1, and in addition suppose $\omega_0 < \omega_1 < \dots < \omega_m$ . Denote by $r$ a constant positive integer, and put

$$\psi(y) = (y-\omega_0)^{2r}\{(y-\omega_1)\dots(y-\omega_m)\}^2 .$$

Let $\mu$ be the maximum of $\psi(y)$ in the interval

$$\omega_0 \le y \le \omega_m ,$$

and let $\mu_k$ , for $k = 1, 2, \dots, m$ , be the maximum of the same function in the subinterval

$$\omega_{k-1} \le y \le \omega_k .$$

Denote by $\varepsilon$ a positive number less than $\mu/2$ . Provided that $r$ is sufficiently large, we evidently have

$$\mu_m = \mu , \text{ but } \mu_k \le \mu-2\varepsilon \text{ for } k = 1, 2, \dots, m-1 .$$

Therefore for all positive integers $\nu$ , by a trivial estimate,

$$R_{k0}\binom{\omega_0,\omega_1,\ldots,\omega_m}{2r\nu,2\nu,\ldots,2\nu} = e^{\omega_0+\omega_k}\int_{\omega_0}^{\omega_k} e^{-y}\psi(y)^\nu dy = O\big((\mu-2\varepsilon)^\nu\big) \qquad (k = 0,1,\ldots,m-1)$$

while on the other hand, by the lemma,

$$R_{m0}\binom{\omega_0,\omega_1,\ldots,\omega_m}{2r\nu,2\nu,\ldots,2\nu} = e^{\omega_0+\omega_m}\int_{\omega_0}^{\omega_m} e^{-y}\psi(y)^\nu dy \geq (\mu-\varepsilon)^\nu$$

for all sufficiently large $\nu$ .

**30.** A slightly different approach is necessary if Späth's method is to lead to the General Theorem of Lindemann.

Assume that $m = p + 2$ and that the equation to be disproved has the form

(36): $$o_0e^{\omega_0} + o_1e^{\omega_1} + \ldots + o_pe^{\omega_p} = 0$$

where the $\omega_k$ and $o_k$ satisfy the conditions (a) and (b) of Hypothesis 2. The terms of this equation may be renumbered, and we also may add an arbitrary positive integer to all the exponents $\omega_k$ without violating the conditions (a) and (b). Hence, *without loss of generality, the real parts of the exponents satisfy the inequalities*

$$0 \leq R(\omega_0) \leq R(\omega_1) \leq \ldots \leq R(\omega_p) .$$

We may in addition, likewise *without loss of generality*, assume that *both* $\omega_0$ *and* $\omega_p$ *are real*. For if $\alpha$ is any algebraic number, then its complex conjugate $\bar{\alpha}$ occurs among the algebraic conjugates of $\alpha$ , and both $\alpha$ and $\bar{\alpha}$ have the same real part and $\alpha + \bar{\alpha}$ is real. It follows that

$$0 \leq R(\bar{\omega}_0) \leq R(\bar{\omega}_1) \leq \ldots \leq R(\bar{\omega}_p) ,$$

and, by the property (b) of Hypothesis 2, that the equation (36) differs from

(37): $$o_0e^{\bar{\omega}_0} + o_1e^{\bar{\omega}_1} + \ldots + o_pe^{\bar{\omega}_p} = 0$$

only in the order of the terms. On squaring the equation (36), or what is the same, multiplying it by the equation (37), we obtain a new equation of the form (36) which still has the properties (a) and (b) of Hypothesis 2, but in which now the first and the last exponent are real, and the real parts of the exponents are non-negative and arranged in increasing order. Again without loss of generality, let this already be the original equation (36). Thus we have now

$$0 \le \omega_0 \le R(\omega_1) \le R(\omega_2) \le \ldots \le R(\omega_{p-1}) \le \omega_p ,$$

and $\omega_0, \omega_1, \ldots, \omega_p$ still are all distinct. For if two or more of the sums $\omega_k + \omega_l$ are equal, the corresponding terms $c_k c_l e^{\omega_k+\omega_l}$ in the product equation can be combined into a single new term without upsetting the conditions (a) and (b) of the hypothesis.

Späth's method can now be applied to this simplified equation (36) in the following way. Denote by $q$, $r$, and $\nu$ three positive integers, put

$$\omega_{p+1} = +\sqrt{-q} , \quad \omega_{p+2} = -\sqrt{-q} ,$$

and define the $\rho$'s by

$$\rho_0 = \rho_1 = \ldots = \rho_p = 2\nu , \quad \rho_{p+1} = \rho_{p+2} = 2r\nu .$$

Hence, on putting

$$\psi(y) = \prod_{k=0}^{p} (y-\omega_k)^2 . (y^2+q)^{2r} ,$$

the functions $R_{k0}$ are given by

$$R_{k0}\begin{pmatrix}\omega_0,\omega_1,\ldots,\omega_p,\omega_{p+1},\omega_{p+2} \\ 2\nu,2\nu,\ldots,2\nu,\ 2r\nu,\ 2r\nu\end{pmatrix} = e^{\omega_0+\omega_k}\int_{\omega_0}^{\omega_k} e^{-y}\psi(y)^{\nu}dy \qquad (k = 1,2,\ldots,p) .$$

Here the integration may be extended over the line segment $L_k$ in the complex plane which connects $\omega_0$ to $\omega_k$.

The two numbers $\omega_0$ and $\omega_p$ by our construction are real, and all the $\omega_k$ are numbered in order of increasing real part. There is then a largest suffix $s$ satisfying $0 \le s \le p-1$ such that $\omega_s$ is real. Denote by $L$ the line segment on the real axis which joins $\omega_s$ to $\omega_p$ ; evidently $L$ is a subsegment of $L_p$.

The function $\psi(y)$ is real and non-negative on $L_p$ and hence also on its subset $L$. Denote by $\mu$ the maximum of $\psi(y)$ on $L$, and by $\mu_k$, for $k = 1, 2, \ldots, p$, the maximum of $\psi(y)$ on $L_k$; let further $\varepsilon$ be a positive number less than $\mu/2$. If $q$ and $r$ are chosen sufficiently large, then evidently

$$\mu = \mu_p \quad \text{and} \quad \mu_k \le \mu-2\varepsilon \quad \text{for} \quad k = 1, 2, \ldots, p-1 .$$

Hence, by a trivial estimate,

$$R_{k0}\begin{pmatrix}\omega_0,\omega_1,\ldots,\omega_p,\omega_{p+1},\omega_{p+2}\\ 2\nu,2\nu,\ldots,2\nu,\ 2r\nu,\ 2r\nu\end{pmatrix} = O\big((\mu-2\varepsilon)^\nu\big) \qquad (k = 1,2,\ldots,p-1)$$

and by the lemma,

$$R_{p0}\begin{pmatrix}\omega_0,\omega_1,\ldots,\omega_p,\omega_{p+1},\omega_{p+2}\\ 2\nu,2\nu,\ldots,2\nu,\ 2r\nu,\ 2r\nu\end{pmatrix} \geq (\mu-\varepsilon)^\nu \ .$$

The sum

$$T(2\nu,2\nu,\ldots,2\nu,2r\nu,2r\nu) = e^{-\omega_0}\sum_{k=1}^{p} R_{k0}\begin{pmatrix}\omega_0,\omega_1,\ldots,\omega_p,\omega_{p+1},\omega_{p+2}\\ 2\nu,2\nu,\ldots,2\nu,\ 2r\nu,\ 2r\nu\end{pmatrix} o_k$$

thus satisfies the inequality

$$|T(2\nu,2\nu,\ldots,2\nu,2r\nu,2r\nu)| \geq e^{\omega_0}\big(|o_p|(\mu-\varepsilon)^\nu - O\big((\mu-2\varepsilon)^\nu\big)\big) > 0$$

as soon as $\nu$ is sufficiently large because $o_p \neq 0$ . This once again proves the assertion.

**31.** There are a number of further classical proofs in the literature, but mostly they are in journals difficult to obtain and will therefore not be discussed. References to these papers can be found in the report on Diophantine Approximations by J.F. Koksma (1936), p. 60.

All these classical proofs are based on Hermite's approximations of the exponential function of §4. Hermite (1893) studied also a second kind of approximations for the exponential function which can likewise be used to derive the theorems of Hermite and Lindemann. Surprisingly, Hermite never applied them to this purpose, and seemingly this was first done by K. Mahler (1931 and 1967). This work by Mahler was much influenced by Siegel's paper of 1929 which by then had appeared and which, as we have seen in earlier chapters, was also basic for the general investigations by Shidlovski.

About the connections between the two kinds of approximations by Hermite see Mahler's paper of 1931. Both kinds of approximations are special cases of the theory of perfect systems; see in particular Mahler (1968).

## REFERENCES


A. Baker (1966), "Linear forms in the logarithms of algebraic numbers I", *Mathematika* **13**, 204-216.

A. Baker (1967a), "Linear forms in the logarithms of algebraic numbers II", *Mathematika* **14**, 102-107.

A. Baker (1967b), "Linear forms in the logarithms of algebraic numbers III", *Mathematika* **14**, 220-228.

Alan Baker (1975), *Transcendental Number Theory* (Cambridge University Press, Cambridge, 1975).

И.И. Белогривов [I.I. Belogrivov] (1967), "О трансцендентности и алгебраической независимости значений некоторых гипергеометрических $E$-функций" [The transcendentality and algebraic independence of the values of certain hypergeometric $E$-functions], *Dokl. Akad. Nauk SSSR* **174**, 267-270; *Soviet Math. Dokl.* **8**, 610-613. See also: *Vestnik Moskov. Univ. Ser. I Mat. Meh.* **22**, no 2, 55-62.

H.F. Blichfeldt (1917), "A further reduction of the known maximum limit to the least value of quadratic forms", *Bull. Amer. Math. Soc.* **23**, 401-402.

P.E. Böhmer (1927), "Über die Transzendenz gewisser dyadischer Brüche", *Math. Ann.* **96**, 367-377; 735.

[Georg] Cantor (1874), "Ueber eine Eigenschaft des Inbegriffs aller reellen algebraischen Zahlen", *J. reine angew. Math.* **77**, 258-262. See also: *Gesammelte Abhandlungen Mathematischen und Philosophischen Inhalts*, 115-118 (Julius Springer, Berlin, 1932; reprinted: George Olms Verlagsbuchhandlung, Hildesheim, 1962).

F.J. Dyson (1947), "The approximation to algebraic numbers by rationals", *Acta Math.* **79**, 225-240.

G. Eisenstein (1852), "Über eine allgemeine Eigenschaft der Reihen-Entwicklungen aller algebraischen Funktionen", *S.-B.K. Preuss. Akad. Wiss. Berlin* (1852), 441-444.

Georg Faber (1904), "Über arithmetische Eigenschaften analytischer Funktionen", *Math. Ann.* **58**, 545-557.

A.O. Гельфонд [A.O. Gel'fond] (1949), "Об алгебраической независимости трансцендентных чисел некоторых классов" [On the algebraic independence of transcendental numbers of certain classes], *Uspehi Mat. Nauk (N.S.)* **4**, no. 5, (33), 14-48; *Amer. Math. Soc. Transl.* (1952), no. 66; reprinted: *Amer. Math. Soc. Transl.* (1) **2** (1962), 125-169.

A.O. Гельфонд (1952), Трансцендентные и алгебраические числа (Gosudarstv. Izdat. Tehn.-Teor. Lit. Moscow);
A.O. Gel'fond, *Transcendental and Algebraic Numbers* (translated by Leo F. Boron; Dover, New York, 1960).

A.O. Гельфонд [A.O. Gel'fond] (1965-1966), "О нулях аналитических функций с заданной арифметикой коэффициентов и представлении чисел" [On zeros of analytic functions with given arithmetic coefficients and representations of numbers], *Acta Arith.* **11**, 97-114.

P. Gordan (1893), "Transcendenz von $e$ und $\pi$ ", *Math. Ann.* **43**, 222-224.

R. Güting (1961), "Approximation of algebraic numbers by algebraic numbers", *Michigan Math. J.* **8**, 149-159.

Ch. Hermite (1873), "Sur la fonction exponentielle", *C.R. Acad. Sci. Paris* **77**, 18-24; 74-79; 226-233; 285-293. See also: *Oevres de Charles Hermite*, III (par Émile Picard), 150-181 (Gauthier-Villars, Paris, 1912).

Ch. Hermite (1893), "Sur la généralisation des fractions continues algébriques", *Ann. Mat. Pura Appl.* (2) **21**, 289-308. See also: *Oevres de Charles Hermite*, IV (par Émile Picard), 357-377 (Gauthier-Villars, Paris, 1917).

David Hilbert (1890), "Ueber die Theorie der algebraischen Formen", *Math. Ann.* **36**, 473-534. See also: "Über die Theorie der algebraischen Formen", *Gesammelte Abhandlungen.* II: *Algebra, Invariantentheorie, Geometrie*, 199-257 (Julius Springer, Berlin, 1933; reprinted: Chelsea, New York, 1965).

David Hilbert (1893), "Ueber die Transcendenz der Zahlen $e$ und $\pi$ ", *Math. Ann.* **43**, 216-219; *Nachr. K. Ges. Wiss. Göttingen* (1893), 113-116. See also: "Über die Transzendenz der Zahlen $e$ und $\pi$ ", *Gesammelte Abhandlungen.* I: *Zahlentheorie*, 1-4 (Julius Springer, Belin, 1932; reprinted: Chelsea, 1965).

A. Hurwitz (1890/91), "Über beständig convergirende Potenzreihen mit rationalen Zahlencoefficienten und vorgeschriebenen Nullstellen", *Acta Math.* **14**, 211-215. See also: "Über beständig konvergierende Potenzreihen mit rationalen Zahlenkoeffizienten und vorgeschriebenen Nullstellen", *Mathematische Werke von Adolf Hurwitz.* I: *Funktionentheorie*, 310-313 (Emil Birkhäuser, Basel, 1932).

A. Hurwitz (1893), "Beweis der Transcendenz der Zahl $e$ ", *Math. Ann.* **43**, 220-221; *Nachr. K. Ges. Wiss. Göttingen* (**1893**), 153-155. See also: "Beweis der Transzendenz der Zahl $e$ " *Mathematische Werke von Adolf Hurwitz. II: Zahlentheorie, Algebra und Geometrie*, 134-135 (Emil Birkhäuser, Basel, 1933).

S. Kakeya (1915), "On an extended Diophantine approximation", *Sci. Rep. Tôhoku Univ.* **4**, 105-109.

J.F. Koksma (1936), *Diophantische Approximationen* (Ergebnisse der Mathematik und ihrer Grenzgebiete, **4**. Julius Springer, Berlin, 1936; reprinted: Chelsea, New York, [1953]; reprinted: Springer-Verlag, Berlin, Heidelberg, New York, 1974).

J.H. Lambert (1770), "Vorläufige Kenntnisse für die, so die Quadratur und Rectification des Circuls suchen", *Beyträge zum Gebrauche der Mathematik und deren Anwendung*, II, 140-169 (5. Abhandlung, Berlin, 1770). See also: Iohannis Henrici Lamberti, *Opera Mathematica*. I: *Commentationes Arithmeticae Algebraicae et Analyticae Pars Prima*, 194-212 (Edidit, Andreas Speiser. Orell Füßli Verlag, Zürich, 1946).

C.G. Lekkerkerker (1949), "On power series with integral coefficients. II", *Nederl. Akad. Wetensch. Proc.* **52**, 1164-1174 = *Indag. Math.* **11**, 438-448.

F. Lindemann (1882), "Ueber die Zahl $\pi$ ", *Math. Ann.* **20**, 213-225.

J. Liouville (1844), "Sur des classes très-étendues de quantités dont la valeur n'est ni rationnelle ni même réductible à des irrationnelles algébriques", *C.R. Acad. Sci. Paris* **18**, 883-885.

Kurt Mahler (1932a), "Zur Approximation der Exponentialfunktion und des Logarithmus. I", *J. reine angew. Math.* **166**, 118-136.

Kurt Mahler (1932b), "Zur Approximation der Exponentialfunktion und des Logarithmus. II", *J. reine angew. Math.* **166**, 137-150.

Kurt Mahler (1937), "Arithmetische Eigenschaften einer Klasse von Dezimalbrüchen", *Akad. Wetensch. Amsterdam Proc.* = *Indag. Math.* **40**, 421-428.

K. Mahler (1965), "Arithmetic properties of lacunary power series with integral coefficients", *J. Austral. Math. Soc.* **5**, 56-64.

K. Mahler (1967a), "Applications of some formulae by Hermite to the approximation of exponentials and logarithms", *Math. Ann.* **168**, 200-227.

К. Малер [K. Mahler] (1967b), "Об одной лемме А.Б. Шидловского" [A lemma of A.B. Šidlovskiĭ], *Mat. Zametki* **2**, 25-32.

K. Mahler (1968a), "Applications of a theorem by A.B. Shidlovski", *Proc. Roy. Soc. London Ser. A* **305**, 149-173.

K. Mahler (1968b), "Perfect systems", *Compositio Math.* **19**, 95-166.

K. Mahler and G. Szekeres (1966/67), "On the approximation of real numbers by roots of integers", *Acta Arith.* **12**, 315-320.

Edmond Maillet (1903), "Sur les séries divergentes et les équations différentielles", *Ann. École Norm.* (3) **20**, 487-518.

E. Maillet (1906), *Introduction à la Théorie des Nombres Transcendants et des Propriétés Arithmétiques des Fonctions* (Gauthier-Villars, Paris).

David Masser (1975), *Elliptic Functions and Transcendence* (Lecture Notes in Mathematics, **437**. Springer-Verlag, Berlin, Heidelberg, New York, 1975).

F. Mertens (1896), "Über die Transzendenz der Zahlen $e$ und $\pi$", *S.-B. math.-nat. Kl. K. Akad. Wiss. Wien II* **105**, 839-855.

Hermann Minkowski (1910), *Geometrie der Zahlen* (Teubner, Berlin; reprinted: Chelsea, New York, 1953).

В.А. Олейников [V.A. Oleĭnikov] (1969), "Об алгебраической независимости значений $E$-функций" [The algebraic independence of the values of $E$-functions], *Mat. Sb. (N.S.)* 78 (**120**), 301-306; *Math. USSR-Sb.* **7** (1970), 293-298.

Oskar Perron (1929), *Die Lehre von den Kettenbrüchen*, 2nd edition (B.G. Teubner's Sammlung von Lehrbüchern auf dem Gebiete der Mathematischen Wissenschaften, **36**. B.G. Teubner, Leipzig, Berlin, 1929; reprinted: Chelsea, New York, 1950).

G. Pólya und G. Szegö (1925), *Aufgaben und Lehrsätze aus der Analysis* (Die Grundlehren der Mathematischen Wissenschaften in Einzeldarstellungen, XIX/XX. Springer, Berlin, 1925; reprinted: Springer-Verlag, Berlin, Heidelberg, New York, 1954).

J. Popken (1929a), "Zur Transzendenz von $e$", *Math. Z.* **29**, 525-541.

J. Popken (1929b), "Zur Transzendenz von $\pi$", *Math. Z.* **29**, 542-548.

Jan Popken (1935), "Über arithmetische Eigenschaften analytischer Funktionen", (Dissertation: Groningen. Noord-Hollandsche Uitgeversmaatschappij, Amsterdam).

K.F. Roth (1955a), "Rational approximations to algebraic numbers", *Mathematika* **2**, 1-20.

K.F. Roth (1955b), "Rational approximations to algebraic numbers: Corrigendum", *Mathematika* **2**, 168.

Theodor Schneider (1934a), "Transzendenzuntersuchungen periodischer Funktionen. I. Transzendenz von Potenzen", *J. reine angew. Math.* **172**, 65-69.

Theodor Schneider (1934b), "Transzendenzuntersuchungen periodischer Funktionen. II. Transzendenzeigenschaften elliptischer Funktionen", *J. reine angew. Math.* **172**, 70-74.

Theodor Schneider (1957), *Einführung in die Transzendenten Zahlen* (Die Grundlehren der Mathematischen Wissenschaften, **81**. Springer-Verlag, Berlin, Göttingen, Heidelberg, 1957).

F. Schottky (1914), "Zu den Beweisen des Lindemannschen Satzes", *Schwarz-Festschr.*, 384-389 (Springer, Berlin, 1914).

K. Shibata (1929), "On the order of the approximation of irrational numbers by rational numbers", *Tôhoku Math. J.* **30**, 22-50.

А.Б. Шидловский [A.B. Šidlovskiĭ] (1959a), "О трансцендентности и алгебраической независимости значений некоторых функций" [Transcendentality and algebraic independence of the values of certain functions], *Trudy Moskov. Mat. Obšč.* **8**, 283-320; *Amer. Math. Soc. Transl.* (2) **27** (1963), 191-230.

А.Б. Шидловский [A.B. Šidlovskiĭ] (1959b), "О критерии алгебраической независимости значений одного класса целых функций" [A criterion for algebraic independence of the values of a class of entire functions], *Izv. Akad. Nauk SSSR Ser. Mat.* **23**, 35-66; *Amer. Math. Soc. Transl.* (2) **22** (1962), 339-370.

А.Б. Шидловский [A.B. Šidlovskiĭ] (1959c), "О трансцендентности и алгебраической независимости значений целых функций некоторых классов" [Transcendence and algebraic independence of the values of entire functions of certain classes], *Moskov. Gos. Univ. Uč. Zap.* **186**, 11-70.

А.Б. Шидловский [A.B. Šidlovskiĭ] (1962), "О трансцендентности и алгебраической независимости значений $E$-функций, связанных любым числом алгебраических уравнений в поле рациональных функций" [Transcendency and algebraic independence of values of $E$-functions related by an arbitrary number of algebraic equations over the field of rational functions], *Izv. Akad. Nauk SSSR Ser. Mat.* **26**, 877-910; *Amer. Math. Soc. Transl.* (2) **50** (1966), 141-177.

А.Б. Шидловский [A.B. Šidlovskiĭ] (1966), "К общей теореме об алгебраической независимости значений $E$-функций" [On a general theorem on the algebraic independence of values of $E$-functions], *Dokl. Akad. Nauk SSSR* **171**, 810-813; *Soviet Math. Dokl.* **7**, 1569-1572.

C.L. Siegel (1929), "Über einige Anwendungen diophantischer Approximationen", *Abh. Preuss. Akad. Wiss. Phys.-mat. Kl. Berlin* (**1929**), no. 1. See also: *Carl Ludwig Siegel Gesammelte Abhandlungen*, I, 209-266 (Springer-Verlag, Berlin, Heidelberg, New York, 1966).

Carl Ludwig Siegel (1949), *Transcendental Numbers* (Annals of Mathematics Studies, **16**. Princeton University Press, Princeton, 1949).

H. Späth (1928), "Zur Transzendenz von $e$ und $\pi$", *Math. Ann.* **98**, 737-744.

В.Г. Спринджук [V.G. Sprindžuk] (1964), "О гипотезе Малера" [On Mahler's conjecture], *Dokl. Akad. Nauk SSSR* **154**, 783-786. See also: *Dokl. Akad. Nauk SSSR* **155**, 54-56.

В.Г. Спринджук [V.G. Sprindžuk] (1967), Проблема Малера в метрической теории чисел [*Mahler's problem in metric number theory*] (Izdat. "Nauka i Tehnika", Minsk, 1967).

Paul Stäckel (1895), "Ueber arithmetische Eigenschaften analytischer Funktionen", *Math. Ann.* **46**, 513-520.

Paul Stäckel (1902), "Arithmetische Eigenschaften analytischer Funktionen", *Acta Math.* **25**, 371-383.

[T.-J.] Stieltjes (1890), "Sur la fonction exponentielle", *C.R. Acad. Sci. Paris* **110**, 267-270. See also: *Oevres Complètes de Thomas Jan Stieltjes*, II, 231-233 (P. Noordhoff, Groningen, 1918).

Kenneth B. Stolarsky (1974), *Algebraic Numbers and Diophantine Approximation* (Pure and Applied Mathematics, **26**. Marcel Dekker, New York, 1974).

Axel Thue (1909), "Über Annäherungswerte algebraischer Zahlen", *J. reine angew. Math.* **135**, 284-305.

Keijo Väänänen (1972), "On a conjecture of Mahler concerning the algebraic independence of the values of some *E*-functions", *Ann. Acad. Sci. Fenn. Ser. A* I **512**.

Keijo Väänänen (1973), "On the transcendence and algebraic independence of the values of certain *E*-functions", *Ann. Acad. Sci. Fenn. Ser. A* I **537**.

O. Venske (1890), "Ueber eine Abänderung des ersten Hermite'schen Beweises für die Transcendenz der Zahl $e$ ", *Nachr. K. Ges. Wiss. Göttingen* **(1890)**, 335-338.

Michael Waldschmidt (1974), *Nombres Transcendants* (Lecture Notes in Mathematics, **402**. Springer-Verlag, Berlin, Heidelberg, New York, 1974).

Heinrich Weber (1899), *Lehrbuch der Algebra*, II, 2nd edition, 822-844 (Vieweg & Sohn, Braunschweig, 1899; 3rd edition: Chelsea, New York, [1961]).

Karl Weierstrass (1885), "Zu Lindemann's Abhandlung 'Über die Ludolph'sche Zahl'", *S.-B.K. Preuss. Akad. Wiss. Berlin* **(1885)**, 1067-1985. See also: *Mathematische Werke*, II, 341-362 (Meyer und Müller, Berlin, 1895; reprinted: George Olms Verlagsbuchhandlung, Hildesheim; Johnson, New York, 1967).